THE
AMERICAN
TRACTOR

A Century of
Legendary Machines

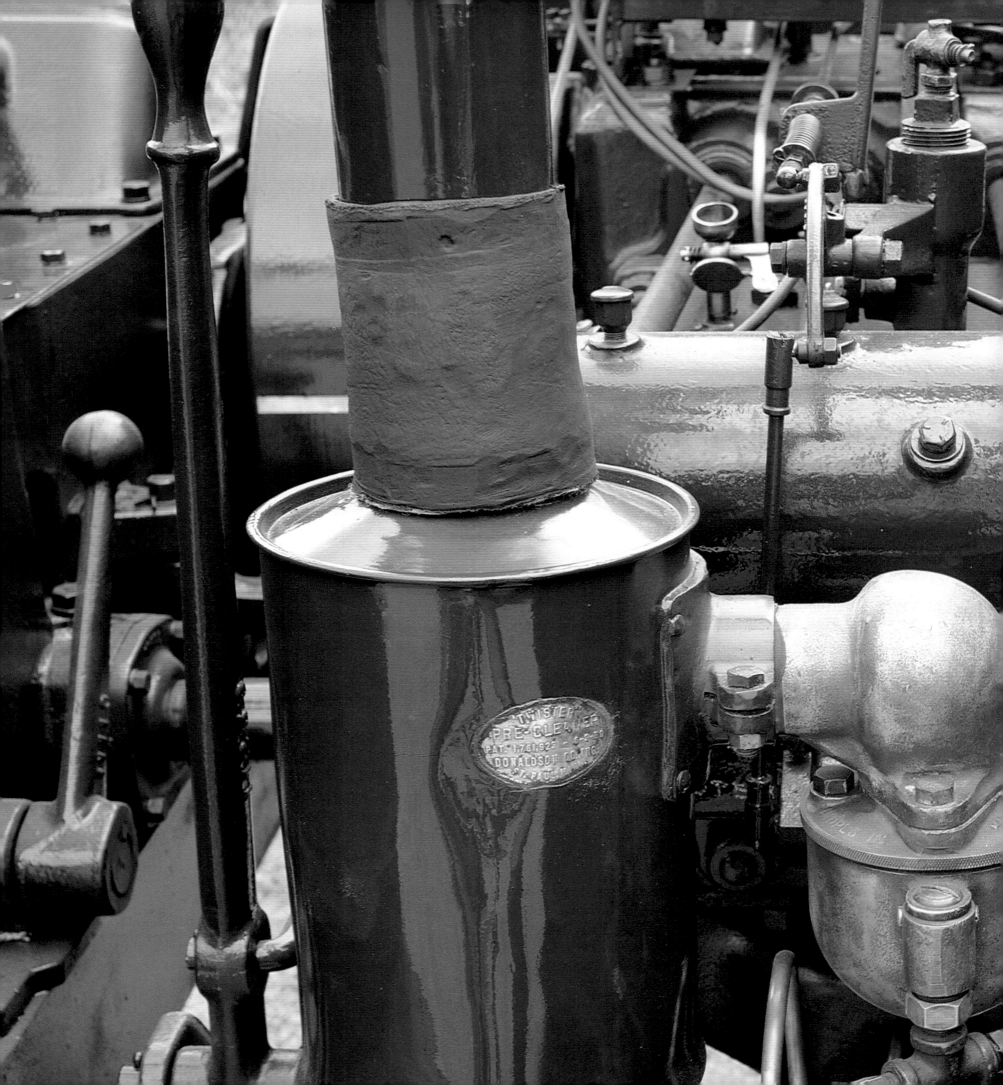

THE AMERICAN TRACTOR

A Century of Legendary Machines

P. W. Ertel

MBI Publishing Company

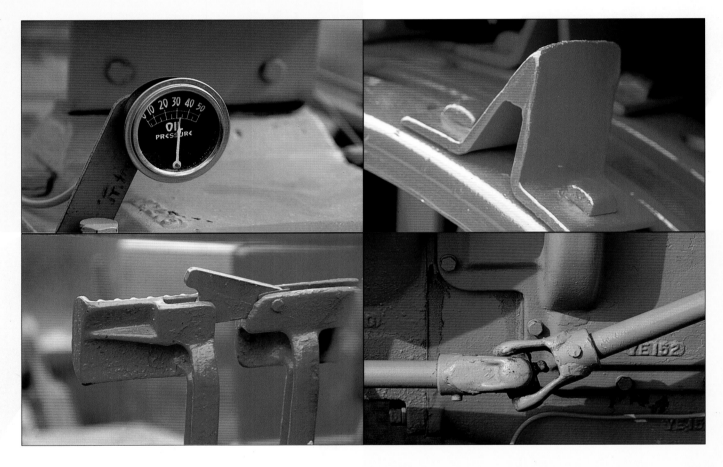

The Author

P.W. Ertel grew up on an Ohio farm in the 1960s, where he acquired an intense interest in farm tractors and machinery. He left the farm to pursue an engineering career, but never lost his appreciation for the machines that had made farm life so enjoyable. He has written about and photographed farm tractors and trucks extensively. He has published one other book on farm equipment, The American Steam Tractor.

Editor **Philip de Ste. Croix**

Designer **Philip Clucas** MSIAD

Landmark photography **Neil Sutherland**

Index **Richard O'Neill**

Publishing manager **Will Steeds**

Production **Neil Randles**

Color reproduction **Studio Technology**

Printed in Spain by Bookprint, S.L., Barcelona

This edition first published in 2001 by MBI Publishing Company, 729 Prospect Avenue, PO Box 1, Osceola, WI 54020-0001 USA

Produced by Salamander Books Ltd, 8, Blenheim Court Brewery Road, London N7 9NT, United Kingdom

A member of the Chrysalis Group plc

© 2001 Salamander Books Ltd

The information in this book is true and complete to the best of our knowledge. All recommendations are made without any guarantee on the part of the author or publisher, who also disclaim any liability incurred in connection with the use of this data or specific details.

We recognize that some words, model names and designations, for example, mentioned herein are the property of the trademark holder. We use them for identification purposes only. This is not an official publication.

MBI Publishing Company books are also available at discounts in bulk quantity for industrial or sales-promotional use. For details write to Special Sales Manager at Motorbooks International Wholesalers & Distributors, 729 Prospect Avenue, PO Box 1, Osceola, WI 54020-0001 USA.

Library of Congress Cataloging-in-Publication Data Available.

ISBN 0-7603-0863-2

CONTENTS

INTRODUCTION

Much has been written about the industrial revolution. But while men were going to work in factories in the cities, there was a quieter revolution taking place on American farms. The back-breaking work and daily drudgery that was farming was changing forever, and the machine that was making the change was the American farm tractor.

The dawn of the 20th century was a time of rapid expansion of the American farm. The free and open lands of the prairies beckoned farmers from as far away as Eastern Europe and thousands of new acres came under the plow every year. But the prairie gave up its riches grudgingly. Power was needed to extract its wealth, and horsepower and manpower proved to be inadequate.

The first mechanical farm power came in the form of steam traction engines. They placed their own demands on the beleaguered farmer. They required huge amounts of water and fuel and a knowledge of the ways of machinery that few in those days possessed. They were adequate for very large farms and for seasonal power in communities of smaller farms, but with their size, expense, and demanding temperament they would never be the daily work partners farmers needed.

Just before the turn of the century the internal combustion engine appeared as a viable alternative to steam. Builders of the

LEFT: International Harvester's 560 was an attempt to cash in on the market for big, powerful tractors in the late 1950s.

first farm tractors mounted the engines on huge heavy machines that were, after a few years of development, adequate replacements for steam traction engines. Their usefulness was still limited primarily to pulling plows in open fields and to providing belt power for threshing machines and sawmills. Large numbers of horses were still needed for the smaller farm tasks and for cultivating row crops. By 1910 it was clear that the internal combustion engine was superior to steam. It was also clear that it opened possibilities for a new kind of machine, one that threatened the horse's place on the American farm.

The demand from farmers for such a machine was at a fever pitch, and the manpower and horsepower needs of World War I fanned the flames. Unscrupulous opportunists flooded the market with tractors accompanied by wild claims of their abilities. They sold thousands of tractors that did little more than enrich their builders. The problem became so bad that in 1920, under the guidance of Wilmot F. Crozier and L.W. Chase, the State of Nebraska passed the Nebraska Tractor Test Bill, establishing a tractor testing station at the University of Nebraska. This objective source of information on the efficacy of each tractor encouraged a confidence in tractors that served both farmers and the tractor industry well.

When Henry Ford introduced his Fordson in 1918 with a selling price far lower than competing machines, he introduced tractor power to ever smaller farms. As the decade of the 1920s dawned, tractors were becoming smaller, more compact, easier to use, more dependable, and less expensive. But farmers were still looking for the replacement for the horse.

The one important farm task that tractors were still unable to do was the cultivation of row crops. In 1924, International Harvester broke through that final barrier when it unveiled the first row-crop tractor, the Farmall. With the Fordson, which was affordable to nearly all farmers, and the Farmall, which allowed one machine to do nearly all the tasks on the farm, the floodgates were opened. Tractors appeared on American farms like crops sprouting in the Spring. The tractor had arrived, and the next 75 years were spent refining it and molding it ever closer to the needs of the American farmer.

From the beginning tractors have been something special to their owners. Memories of what farm life was like before the tractor came along bear fond feelings toward that dependable old companion who eased the daily grind. Perhaps the greatest change in the next century of farming will be that no one will be around who remembers life before tractors. The concept of the tractor as a companion and friend may be lost, and with it will go what made the 20th-century farm tractor unique.

ABOVE: The Ford-Ferguson 9N introduced a new genre to the industry—the utility tractor.

RIGHT: This example of Allis-Chalmers ground-breaking D19 tractor burns liquid petroleum gas.

CHAPTER ONE

THE AMERICAN TRACTOR

★

ALLIS-CHALMERS

U nlike most manufacturers of farm tractors, Allis-Chalmers had no background in agriculture. The Allis-Chalmers Company had been an industrial equipment manufacturer, and as such was dependent for its sales almost entirely on heavy industry and a strong manufacturing economy. In 1913, company president Otto Falk decided the company would diversify by building farm machinery, particularly tractors.

The Early Years Allis-Chalmers' first venture into production of farm equipment was the production of a rotary plow in 1913. Though ten times as large, its function was similar to a modern day roto-tiller. After building a handful, the project was dropped.

The company's next tractor, the Model 10-18, went into production in November 1914. It was a three-wheeled tractor with

ABOVE: Allis-Chalmers hired Harry Merritt to put an end to its failing farm tractor division in 1926. Instead, Merritt applied his remarkable vision and energy to saving it, and within a decade had turned Allis-Chalmers into an industry leader.

Allis-Chalmers tractor engineers ended the decade of the teens by introducing a thoroughly conventional four-wheel tractor of 15-30 horsepower. This tractor was designed so well that in its initial tests it outperformed expectations and was re-rated at 18-30 horsepower. When it was subjected to the rigorous standardized tests at the University of Nebraska in 1920 its horsepower was officially measured at 21 at the drawbar and 33 on the belt.

The 1920s The first project for the new decade was to design a tractor similar in style to the 20-35 to replace the 10-18. This machine, rated at 12-20 and later raised to 15-25, was never a big seller for the company.

The 6-12 continued to sell in small numbers, though the price was eventually dropped to $295. By the mid-1920s, after 10 years of trying, the tractor business was

a single steering wheel in front and two drive wheels at the rear. Though it showed promise, this was not the tractor Otto Falk had in mind. By 1915 an even smaller machine was under development.

The first commercially successful Allis-Chalmers tractor was this tractor, the Model 6-12. This unconventional tractor had two drive wheels in front and a frame that pivoted in the center. The 6-12 sold initially for $850. Fewer that 2000 were produced, but that was respectable by the standards of the fledgling tractor industry.

still not going well for Otto Falk and Allis-Chalmers. The machines were handsome, resplendent in dark green paint with yellow striping and bright red wheels. They were of adequate, if not remarkable, quality and design, but they simply were not selling.

Falk was under pressure to put an end to the tractor experiment. On January 1, 1926 he put Harry Merritt in charge of the tractor division. Merritt cut the price of the 20-35 from $1950 to $1295 and suddenly the tractors were selling. To save the tractor division, Merritt set out to find a

*"Deep plowing—quick plowing—that's the test of a tractor.
It's a race against time and the weather—and your tractor
must deliver tremendous power at high speed!"*

– Allis-Chalmers advertisement 1921

way to produce a tractor of the size and quality of the 20-35 that could be sold at a price the farmer could afford.

Legend has it that Merritt had an entire tractor disassembled and spread over the factory floor. "Every part was scrutinized for ways to make it less expensive to produce. In the end they produced a tractor that weighed 1000 pounds less than its predecessor, was every bit as powerful and sold for a mere $1295." (*An Industrial Heritage*, Walter F. Peterson, p.249). The facts tell a different story. The new

LEFT: Advertising was not enough to sell tractors. Merritt knew a low price and good demonstrations were what sent tractors out the door

ABOVE: The 20-35 was an excellent tractor, but did not sell well because of its relatively high price.

tractor used many of the same parts as the 20-35, and weighed 450 pounds more. Merritt's efforts did allow for the price to be cut to $1495, which helped boost sales from 682 tractors in 1927 to 4867 in 1928. This tractor came to be known as the Model E. This tractor series sold over 16,000 units in 19 years of production.

With its new-found success in agriculture and its background in heavy industry, Allis-Chalmers was well positioned to be a supplier of tracked tractors. Its only problem was it had no such tractor in the inventory. In

ABOVE: Allis Chalmers added crawler tractors to its line by buying the Monarch tractor company in 1928.

The dealers and machinery suppliers who were abandoned by Ford formed the United Tractor and Equipment Distributor's Association. The Association approached Allis-Chalmers with a plan for a new 4000-pound, three-plow tractor to be called the United. The tractor would be built by Allis-Chalmers and sold by members of the Association.

Allis-Chalmers pulled out the stops to design a state-of-the-art tractor and even constructed a modern new factory to build it in. Allis-Chalmers sold the Model U in competition with the United, even though the tractors were identical.

The puzzle of how to make rubber tires work effectively at the heavy drawbar work of general farming had not yet been solved, but Harry Merritt recognized that the day was coming. Even in 1929

ABOVE: The Model U confirmed Allis-Chalmers' recommitment to the tractor business. Also sold as the United, it was a first rate competitor in the crowded three-plow class.

RIGHT: The Model U and its row-crop variant, the Model UC, were powered by a four-cylinder Allis-Chalmers-built engine with a capacity of 301 cubic inches.

1928 Allis-Chalmers purchased the Monarch Tractor Corporation. Its tractor line was obsolete, but it had license agreements that allowed it to use Caterpillar's patents on crawler tracks. After a failed effort to sell the Monarch tractors, Allis-Chalmers devoted the rest of the decade to developing a modern line of crawlers.

Two models, the 35hp Model M and the 62hp Model K were suited to agricultural work. With a lowered seat and full fenders over the tracks, the Model M was popular on orchard and grove farms.

Though the Model E was a solid, well designed, four-plow tractor, the largest market was in two- and three-plow tractors. Several well established manufacturers shared the three-plow tractor market and Henry Ford's Fordson nearly had a stranglehold on the two-plow class. Breaking into the crowded small tractor market would be a daunting, but necessary, task.

Then Ford dropped a bombshell. In December 1927 he announced that the Fordson tractor would no longer be produced in the United States.

truck-type rubber tires could be used on tractors for road transport, and Merritt had the Model U transmission built to take advantage of them. At a time when typical tractors had only two or sometimes three transmission speeds and had a top speed of three or four miles per hour, the Model U came with a 4-speed transmission and was capable of ten miles per hour.

ABOVE AND BELOW: The WC was the first Allis-Chalmers tractor designed from the ground up for row-crop work and the first tractor from anyone to be designed for rubber tires.

The 1930s While other manufacturers had dozens of branch houses and hundreds of dealers throughout the United States, Allis-Chalmers had only five branch houses and a few dozen dealers. The problem of having too few dealers still limited sales. Without a well-developed dealer network, Harry Merritt could never expect Allis-Chalmers to become a power in the tractor industry. Merritt solved that problem in 1931 with the purchase of the Advance Rumely Corporation of La Porte, Indiana. Though Rumely was hopelessly behind in tractor engineering, it had a dealer network that had been developing since the days of the steam engine. With the purchase of Rumely, Allis-Chalmers immediately added about 2500 dealers for its tractors.

Experimentation with rubber tires continued into the new decade. On October 13, 1932 Allis-Chalmers announced that rubber tires would henceforth be supplied as standard equipment on the Model U—a first for an American farm tractor. With its high-speed transmission, the Model U was the perfect machine to lead the charge to equip farm tractors with rubber tires. But 10mph was not enough for the sales department at Allis-Chalmers. They demonstrated rubber tires at ever escalating speeds in tractor races at state fairs across the Midwest, finally reaching 67.877mph on the Utah salt flats.

While Allis-Chalmers was moving ahead of the competition with its lightweight rubber-tired tractors, it was falling behind in the rapidly growing row-crop market. Merritt and his engineers turned

to the Model U as the basis for a tractor they called the Cultivator. Production of this tricycle-type row-crop began in 1930. In 1932 the name was changed to the Model UC. Though it remained in production in small numbers through 1941, the UC was big, heavy, and expensive for a row-crop tractor.

The UC was a good tractor, but it did not quite fit the market as perfectly as a tractor must in order to compete in the hotly contested row-crop market. Allis-Chalmers cleared the slate, started over, and in 1933 introduced the Model WC. The WC was a lightweight two-plow tractor sized perfectly for the small mixed crop operations that made up the majority of family farms in the Midwest. Like any good row-crop tractor,

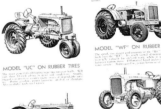

LEFT: From the little 14-horsepower Model B to the 51-horsepower Model A, Allis-Chalmers could supply a row-crop, standard-tread, or crawler tractor to fill any farm need.

BELOW: The massive Model A was built for commercial threshermen. With its large engine and quick governor, it was an excellent power source for sawmills, separators, and all types of belt work.

the WC was versatile, powerful enough to pull a two-bottom plow in nearly any conditions, and nimble enough to cultivate row crops. With its belt pulley and power take-off it could operate any of the machines that required auxiliary power. What made the WC unique was that it was the first tractor ever designed from the ground up to take advantage of rubber tires. What this meant for the farmer was speed. In every gear it was slightly faster than the competition.

In 1936 the Model E was further developed into the Model A. This tractor was built primarily for commercial threshermen. A 4-speed transmission was installed, and the Model E engine was developed into a 510ci powerhouse. These tractors were built in ever decreasing numbers, as the decline of threshing in the late 1930s meant the decline of the big tractor.

The Model B In the early 1930s, Harry Merritt studied the farm census figures and discovered that of the nearly seven million farms in America, some four million were of 100 acres or less. Furthermore, the million or

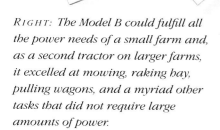

RIGHT: The Model B could fulfill all the power needs of a small farm and, as a second tractor on larger farms, it excelled at mowing, raking hay, pulling wagons, and a myriad other tasks that did not require large amounts of power.

Harry Ferguson had approached Harry Merritt with his hydraulic three-point hitch back in the 1920s, but Merritt wasn't interested. Had Merritt been receptive to Ferguson's new hitch and adopted it for Allis-Chalmers tractors it might well have been the Allis B that introduced the hydraulic revolution to farming instead of the Ford (see pages 54-55).

Over the years the B was upgraded and refined. The first big improvement was the availability of electric starting and lighting in 1940. A simple hydraulic lift soon followed. In 1943 engine modifications gave the tractor a substantial power boost.

After 1950 sales of the Model B began to slide. The B, with its simple 3-speed transmission and without hydraulic draft control, three-point hitch,

so tractors at work on American farms were nearly all on the larger ones. Merritt saw a need for four million small, inexpensive tractors to fill the needs of the small farmers still using horses and set out to build the tractor that would finally put the horse out to pasture.

On December 20, 1937 the revolutionary new Allis-Chalmers Model B was introduced to the world. The chassis design of the B was the result of careful study of its intended market. At 14 horsepower it was obviously meant for the small farm, and the small farms were perched on hillsides and tucked into bottoms, places where the extra stability of the wide front could be a lifesaver.

To allow the B to cultivate row crops, the front axle was steeply arched to raise the frame high, and the frame was made narrow for excellent visibility when cultivating. To make the Model B affordable, it was designed with a minimum of

ABOVE: The RC was an attempt by Allis-Chalmers to build a light tractor that could handle a two-row cultivator.

ABOVE: The Model WF, first introduced in 1937, was a standard-tread version of the popular lightweight WC tractor.

parts and only weighed a ton. The price was just $495, complete with such luxuries as rubber tires and a cushioned seat. Most of all the B was designed to be adaptable. All that was needed was a little ingenuity and the right attachments.

or live PTO, could not compete with newer designs. The tractor continued to be popular overseas, and a diesel engine option was being developed in Europe, but the 1937 design was finally showing its age. The B continued to be built in diminishing numbers through 1957.

1940
Allis-Chalmers Model B

With over four million farms of 100 acres or less, horse farming was still common in North American in the early 1930s. The typical farm of this size could not support a tractor and all the attachments that went with it. The Model B was Allis-Chalmers' attempt to serve this market. With a 1937 price of only $495, it was the answer to many a farmer's dream. While special attachments were available, the B could operate horse-drawn equipment and allowed the farmer to ease into tractor farming, making his investment a little at a time. The really valuable companion piece to the B was the All-Crop Model 40 combine. The All-Crop allowed even the smallest of farmers to harvest his own small grain crops and freed him from the expense of the thresherman. While the Model B failed to put the horse completely out of business, it signaled the beginning of the end for the custom thresherman.

By Allis-Chalmers' standards, the Model B was comfortable. Steering was light, but the hand-operated brakes and forward-mounted throttle were inconvenient.

BELOW: The Model B had a modern engine with full pressure oiling and a spin-on oil filter.

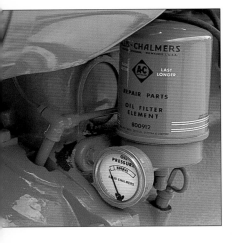

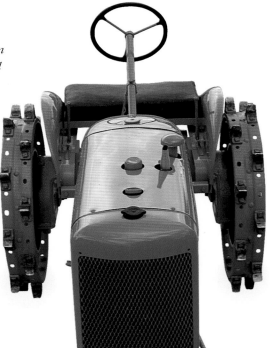

RIGHT: The arched front axle and rear reduction boxes gave the Model B plenty of ground clearance for cultivating a single row.

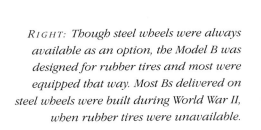

RIGHT: Though steel wheels were always available as an option, the Model B was designed for rubber tires and most were equipped that way. Most Bs delivered on steel wheels were built during World War II, when rubber tires were unavailable.

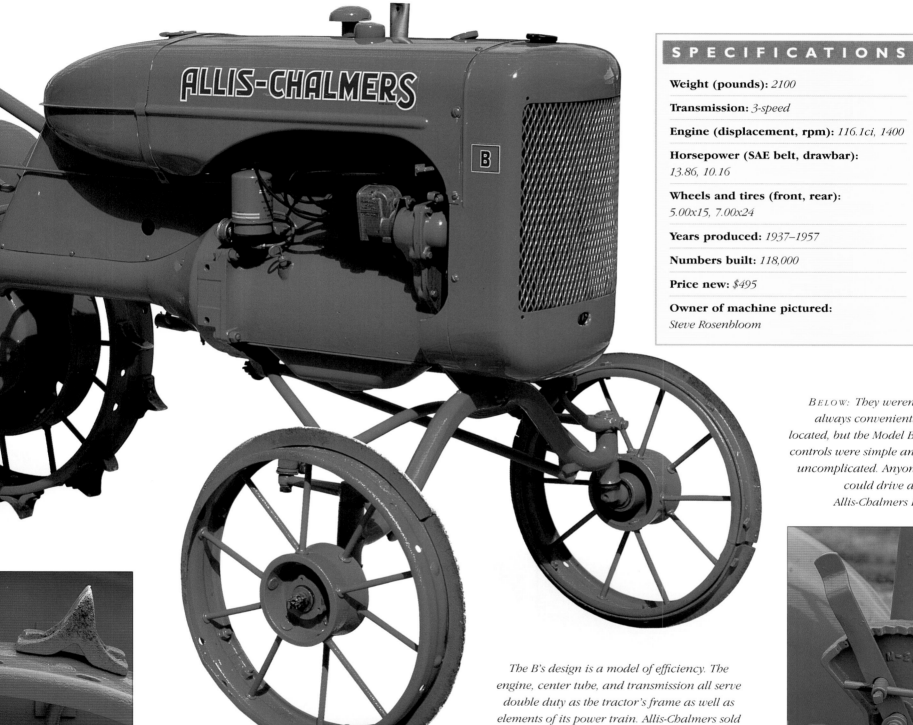

SPECIFICATIONS

Weight (pounds): *2100*

Transmission: *3-speed*

Engine (displacement, rpm): *116.1ci, 1400*

Horsepower (SAE belt, drawbar):
13.86, 10.16

Wheels and tires (front, rear):
5.00x15, 7.00x24

Years produced: *1937–1957*

Numbers built: *118,000*

Price new: *$495*

Owner of machine pictured:
Steve Rosenbloom

BELOW: They weren't always conveniently located, but the Model B's controls were simple and uncomplicated. Anyone could drive an Allis-Chalmers B.

The B's design is a model of efficiency. The engine, center tube, and transmission all serve double duty as the tractor's frame as well as elements of its power train. Allis-Chalmers sold the B for as little as $495. Fully equipped with lights muffler, radiator shutter, and oversize tires, it still cost only $570.

Advertised as "the successor to the horse,"
the Model B was all that and more.

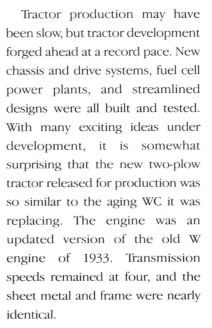

In the late 1930s Allis-Chalmers filled out its line with tractors for more specialized purposes that were adapted from existing models. Several versions of the IB variant of the Model B were available. The moderately popular WF was a standard-tread version of the WC. A Model RC was an ill-fated design that combined a WC chassis and a somewhat more powerful version of the Model B engine. It lasted three model years before being replaced by a properly designed tractor, the Model C.

ABOVE: *A testament to the creative and functional design of the Model G was the rush by specialty tractor manufacturers to copy the design after it went out of production at Allis-Chalmers.*

The 1940s The tractor division of Allis-Chalmers was heavily involved in developing machines to cope with the growing war in Europe, but it did find time to introduce a new tractor, the Model C. The Model C took the same engine as the Model RC and combined it with a well-designed row-crop chassis. The C was the tractor the RC was intended to be; a light two-plow row-crop machine that filled the power gap between the Model B and the WC. The RC was dropped from the line, and the Model C went on to become a very popular model, selling over 80,000 units in its ten years.

Tractor production was severely curtailed for 1943–46, even though the efficiency of the Allis-Chalmers designs allowed the company to build more tractors with less iron and steel than other brands. The last Model A, the direct descendant of the 15-30, first designed in 1918, was built in 1942.

Tractor production may have been slow, but tractor development forged ahead at a record pace. New chassis and drive systems, fuel cell power plants, and streamlined designs were all built and tested. With many exciting ideas under development, it is somewhat surprising that the new two-plow tractor released for production was so similar to the aging WC it was replacing. The engine was an updated version of the old W engine of 1933. Transmission speeds remained at four, and the sheet metal and frame were nearly identical.

Three new features did debut on the WD, one of which was to spread throughout the tractor industry. Allis-Chalmers engineers had developed a round hay baler that required the tractor to stop its forward motion while continuing to power the baler through the PTO. The engineers designed a second, heavy-duty clutch mounted in the power train downstream from the normal flywheel clutch. It effectively provided live power take-off and live hydraulics for the tractor. The hydraulics were needed for the new Traction Booster system, essentially an upside-down version of Harry Ferguson's three-point hitch. The Traction Booster allowed the 4050-pound WD to put its entire 30 horsepower on the ground with minimum wheel slippage.

While neither of these innovations made much of an impression on the industry, the third, power adjust wheels, would be licensed by many manufacturers. Adjusting the width of the rear wheels for the various tasks on the farm had always been a hateful chore. The tractor had to be jacked up and the heavy wheels manually slid on the axles, or worse, removed and reversed. With power

RIGHT: The WD-45 packed a lot of power into a small package.

Powerful WD-45

a NEW HIGH in 4-PLOW POWER

FOR BETTER LIVING
BETTER FARMING
MORE PROFIT . . . *turn the page* . . .

LEFT: With spin–out wheels, hydraulics, and a kind of live power take-off, the WD kept Allis-Chalmers abreast of the postwar advances in tractor technology.

adjust wheels, the tractor did not even need to be raised off the ground. Allis-Chalmers engineers had turned a back-breaking, time-consuming chore into a job that anyone could do in a few minutes.

One of the ideas experimented with during the war took shape in 1948 as the unorthodox Model G. Weighing only 1300 pounds, with a little 10hp Continental engine hung off the back and not much of anything up front, the Model G was a radical departure from standard tractor design. Looking like a motorized insect, it broke the mold for what a farm tractor should look like, but farmers took to it like, well, like ducks on a June bug. Just under 30,000 Model Gs were produced between 1948 and the end of production in 1955.

The 1950s The 1940s were a decade filled with experimentation, but not many new products. The 1950s were to change all that. An improved Model C, the CA, was introduced in late 1950. As soon as dealers and farmers saw the new features on the WD, they wanted them applied to the Model C also. With Traction Booster, a transmission clutch, and power adjust wheels, the CA was the tractor they asked for. The transmission clutch in the CA was a design not worthy of Allis-Chalmers engineers. A clutch was placed in the right rear axle housing which, when disengaged, removed power from the right rear wheel and set the differential spinning madly as long as the clutch was disengaged.

There was little wrong with the WD that a little more horsepower wouldn't improve. In mid-1953 Allis-Chalmers engineers put the power in. The old Model W engine was given increases in displacement and compression and a modified combustion chamber. The results were dramatic. Power was boosted from 35 to 43 horsepower—a 25 per cent increase in power for no increase in weight. The tractor was called the WD-45. An LP fuel version soon followed, and in 1954 a six-cylinder diesel version was introduced. The WD-45, this incredibly light, powerful new tractor that looked like a 1938 WC, caused a sensation.

In 1957 Allis-Chalmers began a program of modernizing its tractors. The first new tractor was the 30hp D14, introduced in the Spring; the 46hp D17 followed it in the Fall. The new tractors featured many innovations. The driver's seat was moved forward, ahead of the rear axle, to give a more comfortable ride. The old hand clutch of the WD was

1963
Allis-Chalmers D19 Turbo

Turbocharging had been in use in diesel engines for some time, but Allis-Chalmers was the first tractor manufacturer to adapt the technology to its farm tractor engines. The North American tractor industry was engaged in a horsepower race that rivaled the arms race in intensity. Any technique that could be used to squeeze more power out of existing designs was employed. Allis-Chalmers stepped out in front of the rest of the industry, as well as avoiding significant additional tooling expense, when it adapted its existing 262 cubic inch gas engine to diesel and then turbocharged it. Turbocharging was necessary to bring the diesel up to near the horsepower level of the gas version. The D19 caused a sensation in the industry and among farmers. It paved the way for other manufacturers, and within a few years every major tractor manufacturer had a turbocharged diesel in the line.

As the rear area of tractors became cluttered with equipment, it was no longer a convenient way to mount the machine. Allis-Chalmers moved the operator station forward, and allowed the operator to mount from the front.

Unlike their predecessors, D series tractors were comfortable and easy to operate.

ABOVE: Allis-Chalmers adopted a clean, no-nonsense look for its tractors. The simple grille was easy to remove and clean.

SPECIFICATIONS

Weight (pounds): *6570*

Transmission: *8-speed*

Engine (displacement, rpm): *262ci, 2000*

Horsepower (belt, drawbar): *66.92, 62.05*

Wheels and tires (front, rear): *6.50x16, 15.5x38*

Years produced: *1961–1964*

Numbers built: *10,597*

Price new: *$5834*

Owner of machine pictured:
Gail Schipansky

ABOVE RIGHT: Allis-Chalmers was the first to offer power adjustable, "spin-out" rear wheels. Spin-out wheels were a $66.95 option on the D19 and were a real time saver for busy farmers.

RIGHT: The D17 used the same 262ci engine without the turbo-charger and produced only 51.14 horsepower.

The D19 was available with either Allis-Chalmers' proprietary Snap-Coupler hitch or a standard category 2 three-point hitch. Both incorporated Allis-Chalmer's Traction Booster – a draft sensing system that transferred weight from the implement to the tractor under heavy draft conditions.

ABOVE: Gauges and controls were all conveniently positioned. A tachometer was included for precise control of power take-off speeds.

LEFT: The Power-Director allowed no-clutch shifting between two adjacent speeds.

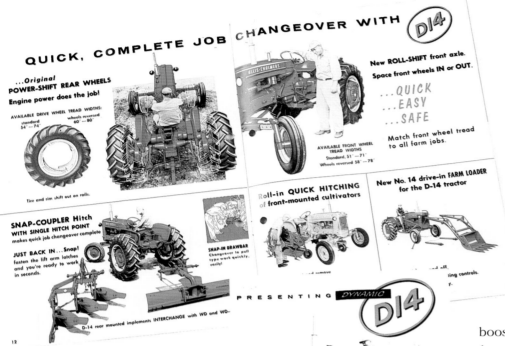

ABOVE AND RIGHT: The D14 could be equipped with Power-Shift rear wheels and Roll-Shift front axle to ease the job of adjusting wheel width. In the 1950s Allis-Chalmers recognized that operator comfort and convenience increased productivity, efficiency, and profits.

The most exciting development was the D19, introduced in 1961. At 71 horsepower, the gas version was the most powerful tractor ever for Allis-Chalmers. Though the diesel engine produced only 67 horsepower, it was the cause of all the excitement. The D19 was the world's first turbocharged diesel farm tractor. Allis-Chalmers did not stop there. The new D21 broke the 100 horsepower barrier just two years later. This 10,700-pound tractor pounded out 103 PTO horsepower from a 426ci diesel engine. By 1965 the One-Ninety XT, with its little 301ci turbocharged engine was producing nearly that much power, but weighed over a ton less. Not to be outdone, the D21 was given a turbocharger, which boosted horsepower to 127. Allis-Chalmers had launched a horsepower war—with itself!

The remainder of the decade was given to consolidating the many old designs under one style of sheet metal and a consistent naming convention, the One Hundred series. The D21 was the last holdout, finally sharing the styling of the rest of the line and given the name Two-Twenty.

retained, but now incorporated a dual range feature that could be shifted from high to low range without stopping the tractor. The D14 received a new 149ci four-cylinder engine. Allis-Chalmers engineers went back to the well once again and put a newly uprated version of the venerable old Model W engine in the D17.

1959 saw the introduction of the little D10 and D12. These tractors, both with updated versions of the 1938 Model B engine, filled the space left by the demise of the Models B and CA. They differed only in that the D12 had a wider tread.

The 1960s The innovative new tractors of the late 1950s were an indication of things to come in the next decade. Power was the word at Allis-Chalmers for the 1960s. The most powerful tractor in the line in 1960 was the 46hp D17. In 1969 A-C's most powerful tractor cranked out 136 horsepower.

ABOVE: With a 139ci engine that was a direct descendant of the engine in the 1938 Model B, the D10 was the smallest Allis-Chalmers tractor of its day.

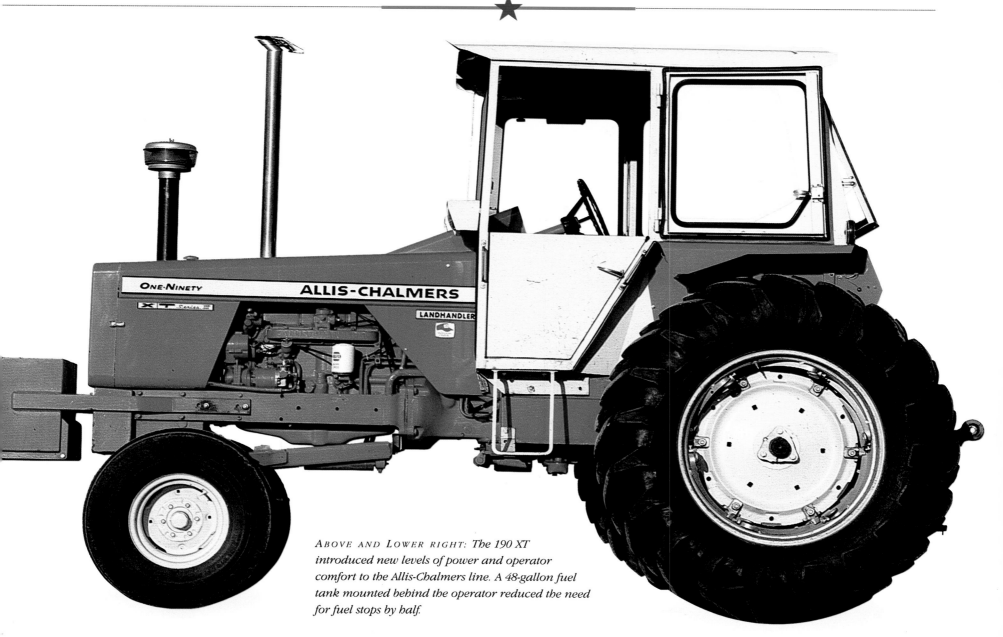

*ABOVE AND LOWER RIGHT: The 190 XT
introduced new levels of power and operator
comfort to the Allis-Chalmers line. A 48-gallon fuel
tank mounted behind the operator reduced the need
for fuel stops by half.*

*ABOVE: In 1960 the D15 introduced a new shade of the famous
Persian orange that Allis-Chalmers had used since 1929.*

The 1970s Allis-Chalmers realized in the 1960s that it could not build small and mid-range tractors profitably in the United States. In the 1970s, the company concentrated on designing new tractors of up to 250 horsepower and purchasing its smaller tractors from overseas manufacturers. The imported tractors were given Allis-Chalmers

sheet metal and paint and sold as Allis-Chalmers tractors. While this helped the company sell new machines, having one line of tractors designed and built on three continents did little to promote efficient service and parts supply.

*Allis-Chalmers' Acousta Cabs
set industry records for providing
a quiet operator environment.*

RIGHT: Unable to build small tractors profitably in the U.S. in the late 1960s, Allis-Chalmers began importing them from France, Romania, and Italy. Imports from Toyosha of Japan, such as this three-plow 6140 proved especially favorable and were part of the A-C line for over a decade.

Allis-Chalmers had relied on its fine crawlers to fill the growing demand for 150hp tractors. But the Construction Machinery Division took over crawler sales, leaving farm equipment dealers without a big tractor to sell. In 1972 Allis-Chalmers filled the gap in its line with Bearcat tractors bought from Steiger. Finally, in 1976 Allis-Chalmers built its first four-wheel-drive tractor, the 23,500-pound, 186hp 7580. Within a year the 253hp 8550 took its place at the top of the A-C line. Though A-C was a late comer to the four-wheel-drive trade, its tractors were equal to the best in the business and sold well.

ABOVE: AGCO was formed by some of Deutz-Allis's management to purchase Deutz-Allis from its parent, KDH.

The 1980s The dawn of the new decade saw the company offer a line of 13 models ranging from 21 to 253 horsepower. In spite of its aggressive engineering, Allis-Chalmers had seen its market share drop steadily throughout the 1970s. A long period of low crop prices and high interest rates initiated an industry-wide decline in tractor sales in the 1980s. Only the strong would survive this devastating descent. Allis-Chalmers was not one of the strong. On March 29, 1985 the *Milwaukee Sentinel* reported that the Allis-Chalmers farm equipment business was to be sold to Klockner-Humbolt-Deutz AG of Germany for $100 million. KDH did

ABOVE: AGCO relied heavily on nostalgia in its marketing. The return to Persian orange (from Deutz-Allis green) was a component of this strategy.

BELOW: The 9815 abandoned the imported air-cooled engine of the Deutz-Allis era and replaced it with a domestic Detroit diesel.

LEFT: For its first line of tractors, including the 6690, AGCO-Allis turned to European tractor manufacturer SAME-Lamborghini-Hurliman. The S+L+H tractors made up the small and medium tractors in the line, while larger tractors were built in the U.S.

ABOVE: With 120 PTO horsepower, the 8630 was powered by a 366ci, turbocharged, air-cooled, six-cylinder diesel engine. A 36-speed transmission and front-wheel-assist were available.

not buy the plant where Allis-Chalmers tractors were made, having no intention of building tractors in the United States. The new company purchased 6000, 8000, and 4W305 model tractors from the plant through early December 1985. On December 6, 1985 the last tractor to be built in the plant, a Model 6070, rolled off the line. This series of purchased tractors and all tractors in stock at the time of the purchase were re-badged with the name Deutz-Allis. No tractors with Allis-Chalmers content were produced by Deutz-Allis after December 6, 1985.

But Deutz had less luck in the U.S. marketplace than had Allis-Chalmers. In 1990 KHD sold its Deutz-Allis division to AGCO Corporation, headquartered in Duluth, Georgia. The Allis name and familiar orange paint returned in 1990 when AGCO began manufacturing and distributing farm equipment under the AGCO-Allis name. Nine models of tractors ranging from 45 to 145 horsepower are marketed under the AGCO-Allis name. AGCO has become a worldwide farm machinery company by acquisition of existing companies and enters the new millennium as one of the world's leading manufacturers of farm tractors.

CHAPTER TWO

THE AMERICAN TRACTOR

★

J.I. CASE

At the turn of the century the name Case was synonymous with farm power. Case had been the pre-eminent builder of agricultural steam engines for 20 years, and had built experimental gasoline tractors as early as 1894. Naturally the public looked to Case for leadership in gas tractor design, leadership which emerged in 1912.

The first commercially viable Case gas tractor was the Model 60, introduced in 1912. The 60 weighed over 25,000 pounds and produced 30 horsepower at the drawbar, 60 at the belt. Over 350 were built in its four years of production. A smaller tractor of 20/40 horsepower was also introduced in 1913. Weighing in at 13,000 pounds, the "little" Case sold over 4200 units before being discontinued in 1919.

Case was considered a conservative company, building machines of weight and substance, but without having much to offer in terms of imagination. This may have been true of steam engines, but once tractor production began, experimentation with new designs was continuous. Case introduced eight new tractor designs in its first seven years.

Two major problems with early tractor designs were the primitive engines and the flexibility of the commonly used riveted frame chassis. If the engines held up long enough, the flexing frame allowed the transmission gears to shift out of alignment, bind, and fail.

Setting its sights directly on these reliability problems, Case introduced a tractor in 1917 that would set the standard for tractors for the next

ABOVE: Case's Crossmotor design allowed for a rigid frame that greatly increased tractor life.

decade. The new 10-18 featured a rigid cast-steel frame that enclosed and supported the transmission gears, and a water washer type air filter that gave the engine longer life. The tractor quickly set new standards for reliability.

Immediately following the 10-18 came the legendary 15-27. Essentially a scaled-up 10-18, the 15-27 had a four-cylinder 382ci engine with pressure oiling. It weighed 5700 pounds, could pull three plows all day long, and cost a reasonable $1680. The 15-27 outsold all other tractors of its size. The engine was designed to start and warm up on gasoline then run on kerosene. Case advertised its tractors as the simplest design possible, both in operation and maintenance. Most tractor buyers in those days were first time owners, and Case

Crossmotor tractors were indeed an excellent choice for a farmer with no tractor experience.

Its power and compact size allowed the 15-27 to enter new markets for Case. The 15-27 was the first Case outfitted with lowered exhaust and intake pipes and special fenders for orchard work. It was a popular industrial tractor, finding work on docks and in factories and yards across the country. A power take-off was available,

Case, fully aware of its standing in the world of steam engines, declared the Crossmotors, "Worthy of the name Case," in its advertising.

which the 15-27 was well designed to make use of, pulling power binders and mowers.

A third Crossmotor, the 22-40, was introduced just in time to usher out the decade. The 22-40 offered all the features of the 15-27 in a heavier, more powerful tractor.

The 1920s Case greeted the 1920s with one of the most massive machines it had ever built—the 40-72. The tractor was similar to the three other Case Crossmotor tractors, the difference being one of scale, but what a difference! Standing eight feet tall at the radiator cap and weighing 21,000 pounds, it was twice as tall and three times as heavy as the 10-18. Only 41 of these tractors were built. Its $4900 price tag was surely a factor, but the small demand for a tractor of this size at any price certainly limited sales numbers.

A second generation of Crossmotor tractors was introduced in the early 1920s, beginning with the 12-20 in 1921. This tractor replaced the 10-18. The engine displacement was increased to 267ci and the spoked wheels were replaced by unique stamped steel wheels. This was the only model of Case tractor to use these wheels. The colorful green, red, and black color scheme was replaced with pinstriped battleship gray paint. Late in 1924 the 15-27 was upgraded to a rating of 18-32 horsepower and the 22-40 became the 25-45. Though experiments were made with cultivator-type tractors, these three Crossmotor models were the extent of the Case tractor line until the introduction of the Model L in 1929.

ABOVE: The Model 22-40 departed from the cast frame design of most Case Crossmotor tractors, utilizing an I-beam frame.

LEFT: Case was a prolific producer of sales material.

ABOVE TOP: The first new tractor of the post-Crossmotor era was the formidable Model L. The Model L launched Case from a trailing position directly to the front in the tractor technology race of the 1920s.

1919
Case 15-27

One of the first truly reliable mid-sized tractors built was the Case 15-27. It was designed along the same lines as the 10-18 that preceded it, but was sized perfectly for a large number of farm operations, and came along at a time when many of the mid-sized gas traction engines were wearing out. The 15-27 was there to take the job. Lighter, faster, and more reliable than the traction engines that preceded it, the 15-27 ushered in a new era of confidence in farm tractors. It had a fully enclosed engine, an effective air washer, and a unit frame, all of which made the 15-27 last hundreds of hours longer than its predecessors. While it sold fewer than 18,000 units in five years, it showed farmers that a tractor didn't need to be cantankerous and expensive and it helped open the door to a new era in power farming.

RIGHT: Mounting the engine across the frame avoided the necessity of turning the power 90 degrees for the rear axle and belt pulley. The design required only simple spur gears. Case gearing was known for its dependability.

The large, slow-turning engine dominates a side view of the tractor. With overhead valves and pressure oiling, Case's engine was quite advanced.

ABOVE: Controls were convenient by the standards of the day; you could reach most of them without getting off the tractor!

ABOVE: The compact, rigid framework of the Case Crossmotors made for rugged tractors. Case Crossmotors, including the 15-27, were renowned for their smooth power. They made great tractors for belt-powered machinery.

SPECIFICATIONS

Weight (pounds): *5700*

Transmission: *2-speed*

Engine (displacement, rpm): *381.7ci, 900*

Horsepower (belt, drawbar): *27, 15*

Wheels and tires: *Steel 32x6, steel 52x14*

Years produced: *1919–1924*

Numbers built: *17,629*

Price new: *$1680*

Owner of machine pictured: *Fred Buckert*

The main fuel tank holds 20 gallons of kerosene and the starting tank holds 2¾ gallons of gasoline.

BELOW: The 15-27 burned kerosene, and had an elaborate manifold for heating the incoming air.

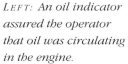

LEFT: An oil indicator assured the operator that oil was circulating in the engine.

RIGHT: This elaborate mechanism is the radiator cap. The cooling system held 11 gallons of water, which was circulated by a centrifugal pump, another advanced Case feature.

BELOW: The Model C was a smaller version of the L.

BELOW: Two more models, the 30-horsepower Model C standard tractor and the Model CC, Case's first row-crop tractor, followed quickly on the heels of the Model L.

As good as the Crossmotor tractors had been at their introduction, they were becoming obsolete by the mid 1920s. In 1925 Case engineers were working on a new tractor to replace the Crossmotors and planned to introduce it to the market before 1930.

The new tractor, called the Model L, was the right tractor at the right time for Case. It was a thoroughly up-to-date machine, with a smooth-running long-stroke engine of 403ci. The Model L weighed just over 5000 pounds and produced 44 belt horsepower. Four or five 14-inch plows were recommended, and at a price of $1295, it was a horsepower bargain.

Late in 1929 a scaled-down version of the Model L was introduced. The two-plow Model C boasted a 260ci four-cylinder engine and weighed only 4000 pounds, but it produced horsepower figures of 30 on the belt and 17 at the drawbar. The Model C was the last new tractor of the decade from Case, but it was to provide the basis for one of their most popular lines.

LEFT: Motor-Lift implements could be changed in a matter of minutes.

The 1930s What would be a slow decade for the Case engineering department began with a flurry of activity over the company's first row-crop tractor. The tractor had been planned for introduction in 1929, but

LEFT: The CC's "chicken roost" steering arm was ungainly, though functional and simple.

BELOW: Case's CC used the same 260ci engine as the Model C.

building a row-crop tractor around the Model C engine proved to be more of a challenge than anticipated. Problems with the steering and front bolster plagued development and the tractor wasn't widely available until 1930. Cultivators, listers, and middle busters were hastily designed for the tractor, and by 1933 a complete general purpose farming system was available. With 18 drawbar horsepower, the CC could handle a four-row cultivator. The CC was a good tractor and was well appreciated by many, but as the only row-crop tractor in the line-up it did little to change the public perception of Case as a builder of big standard-tread tractors.

In 1934 Case salesmen watched with dismay as customers flocked to the inexpensive one-plow Farmall F-12. They wanted a similar tractor. Leon Clausen was adamant that Case could make no money selling a

ABOVE: Case ads praised the Model L's adaptability, dependability, ease of handling, and economy of operation

"Flambeau is the French name for a torch that burns with a reddish light. Flambeaux have lighted the way for explorers, conquerors, pioneers." – Case press release

scaled-down CC. The sales staff didn't need to point out that Case made no money on the F-12s that Farmall was selling by the carload. Clausen eventually relented. The new tractor was the Model RC. With its 132ci four-cylinder engine it produced 14 drawbar and 20 belt horsepower.

The Model D The D series of tractors was introduced in 1939, and with them came a flamboyant new paint scheme, Flambeau Red and black. In the late 1930s every tractor manufacturer rushed to give modern art deco shapes to their machines. Some were more successful than others. With the steering rod extending down the side of the tractor and the gooseneck front wheel support, it was nearly impossible to eliminate the antique look from Case row-crop tractors.

RIGHT: Case president Leon Clausen hated the Model R as much as he hated the RC. He was opposed to the idea of a small row-crop tractor.

BELOW: The RC's cast iron steering wheel was no friend on a cold winter morning.

CASE "RC" *"a Complete"* ALL-PURPOSE TRACTOR

ABOVE: Though early RCs had a more common steering gear, Case soon reverted to its "chicken roost" steering.

RIGHT: An attempt was made to make the RC more stylish, but the overall design of the tractor frustrated attempts at an integrated look..

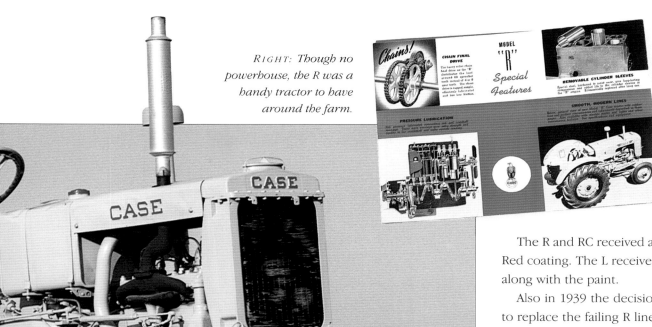

RIGHT: Though no powerhouse, the R was a handy tractor to have around the farm.

Nevertheless, the row-crop version of the D, the DC, was the first effort at styling. This "new" tractor was little more than a CC with rounded sheet metal and red paint. In fact, the first DCs were rebuilt CCs.

The R and RC received a sunburst design grille and the same Flambeau Red coating. The L received an improved cylinder head, new sheet metal along with the paint.

Also in 1939 the decision was made to design an all-new small tractor to replace the failing R line. The result was the Model V and its row-crop and industrial variants, the VC and VI. The V was as hated by Clausen as the R had been. It was relatively light, inexpensive, and had a spur gear final drive—all very unlike traditional Case designs. Soon Clausen had

ABOVE: With its sleek sheet metal and Flambeau Red paint, the Model S at last gave Case a modern-looking tractor. Mechanically it was the same sturdy and dependable design that was introduced on the Model C in 1929.

another two-plow tractor in Case showrooms, and this was one he could embrace. The S series was a scaled-down C and had the low power to weight ratio, high price, and roller chain final drive that Clausen was so fond of. It was heavier and slightly more powerful than the V. Power output from the S was so similar to the V that the

BELOW: The Model V made an excellent platform for an orchard and grove tractor.

ABOVE: Under the stylish sheet metal the LA was fundamentally the same as the Model L it replaced.

ABOVE RIGHT: The Parts List and an open spread from the Operator's Manual for the Model LA.

The 1940s The V was assembled largely from vendor-supplied parts, which made it unprofitable. In late 1941 the first of a redesigned version, the VA came off the production line. The VA was similar to the V, but contained many more Case-built components.

Case was slipping behind technologically by the end of the war. Except for the Model VA every tractor in the line was 20 years old. The LA was mechanically the same as a 1929 L, the D series was the same as the 1929 C, and the S was a scaled-down D. Two postwar developments helped keep

two tractors would have competed for the same market. Clausen's options were to increase the power of the S or limit the V. Given his affinity for heavy, understressed machines, it is no surprise that he had the Model V governor adjusted to limit its power output.

The 1956 Case 400 set a new world's record for fuel economy.

Case in the market. The Eagle Hitch, an easy-to-use type of three-point hitch, offered hydraulic lift but no draft control. A modification of the VAC, called the VAC-14, gave Case a low tractor to compete with the Ford and Fergusons that were sweeping the small tractor field.

The 1950s Case tractors of the early 1950s lacked features such as draft control, live power take-off, and power steering that were becoming common on other makes. Nearly every major manufacturer had diesel engines for their big tractors and many offered diesel power for their larger row-crops. Case had no diesel at all, and this was the area that Case chose to attack first. A strong, versatile new engine, designed from the block up to be an integral part of a whole series of farm tractors, was introduced in 1953. The basic engine had a 4x5 inch bore and stroke. It was designed to use a number of identical two-cylinder heads, so it could be easily and economically built as a four- or six-cylinder engine. What's more, the block was designed to accept heads for diesel, LP, or gasoline fuel. The new engine was first used in the Model 500. Though the 500 was little more than a 1929 Model L with a modern diesel engine, it proved a useful proving ground for the new engine.

The first all-new machine from Case in 26 years was the 400, introduced in 1955. Many of its features came as a result of a questionnaire Case had issued to customers asking what they wanted to see in their next Case tractor. A diesel-powered row-crop, more power, more transmission speeds, and elimination of the "chicken roost" steering rod were high on the list.

The 400 sported all the features of a modern tractor from power steering to attractive paint and styling. Under the hood, it eliminated the roller chain final drive that had been a Case hallmark for decades, introduced the first 8-speed transmission in a row-crop tractor, and was equipped with a thoroughly modern, 49-horsepower diesel engine.

ABOVE: The 400 was available in many configurations, and was a popular high-crop tractor.

LEFT AND BELOW: With the 300 series Case made a substantial step forward in power, comfort, and convenience for the smaller farmer. It offered a long list of engine, transmission, and chassis options.

*"Daring 3-plow Design ...Dazzling New Beauty ...
Dynamic Performance ...The Case 300"*

– Case advertisement 1956

Introduced in 1956, this tractor was sized to fill the 30-horsepower space formerly filled by the Model S. The 300 was available with a 149ci Case gasoline engine or a 157ci diesel engine built by Continental. A 12-speed transmission was available, as was a hand-operated transmission clutch. While the 300 wore the same attractive Desert Sand and Flambeau Red color scheme as the 400, the sheet metal was shaped completely differently. With the 300, 400, and 500 tractors all wrapped in different sheet metal, Case had no brand identification by style, a serious marketing error.

Case added a torque converter option to the 8-speed transmission in its big tractors and called it Case-o-Matic. Case's implementation of the torque converter included a locking feature, giving the operator the option of fluid or direct drive. Case-o-Matic was an inexpensive and effective way to provide the multiple ground speeds farmers were demanding.

ABOVE RIGHT: Case's big wheatland tractor, the Model 600, matched a modern Case diesel engine with a drive train that was a direct descendant of the 1929 Model L. The Case-o-Matic drive (above) allied a torque converter to an 8-speed transmission.

With the exception of the 300 series, which continued unchanged until 1959, Case consolidated its styling and identification system in 1957. A flurry of new models was introduced because Case preferred to identify every engine, transmission, and chassis combination with a new model number. The 801, for example, is a row-crop chassis equipped with a Case-o-Matic transmission. A Model 701 is the same tractor with the standard 8-speed transmission. By the end of the decade Case's agricultural tractor line ranged from the 3600-pound, 30hp 200 series to the aging 70hp, 8500-pound descendant of the Model L, the 910. After being woefully behind the competition at the beginning of the decade, Case designs had surpassed many manufacturer's tractors by 1959.

Farmers prided themselves on their immunity to fatigue and discomfort and weren't likely to buy on that basis, but more money in the pocket—now that was hard to argue with.

The 1960s Forcing two decades of innovation into the latter part of the 1950s took its toll. The company entered the 1960s deeply in debt. It was time for engineering and design to take a backseat to product placement and sales.

Case introduced the 30 series in 1960. The 430, 530 and 630 replaced the 30hp 200 and 300, the 37hp 400, and the 44hp 500 series respectively. The 730 and 830 replaced the 50hp 700 and 800 series tractors, while the 930 was yet another update of the, by this time, ancient 900. No substantial changes were implemented on the new tractors.

In 1962 Case turned to operator comfort as a way to distinguish its tractors from the competition. The operator platform of the 930 was moved forward and isolated from chassis vibration by raising it above the chassis and mounting it on rubber. Case sold its Comfort King concept by impressing upon farmers the economic wisdom of operator comfort— improved comfort means less operator fatigue, more hours in the field, and more dollars in the farmer's pocket. Farmers prided themselves on their immunity to fatigue and discomfort and weren't likely to buy on that basis, but more money in the pocket—now that was hard to argue with.

The Comfort King was a success on the 930, and in 1964 the concept was applied to the 830 and 730 series.

Also in 1964, Case introduced its largest row-crop tractor ever, the 930 GP. This was a marriage of the 730 row-crop chassis and the six-cylinder engine of the 930. At

ABOVE AND TOP RIGHT: New headlights mounted high in the hood gave the early 1960s Case tractors a powerful, aggressive look.

LEFT: Case recognized that operator fatigue was a limiting factor in farm efficiency. The Comfort King cab minimized fatigue and made Case a leader in operator safety and comfort.

85 horsepower and 8800 pounds, the 930 GP was pushing the limits for a two-wheel-drive row-crop tractor.

Early in the decade Case began adapting its farm tractors for use in the construction industry. Backhoes and loaders were designed for the 200 through 500 series of tractors. While it had no direct effect on agricultural tractors, Case became an important supplier to contractors, significantly enhancing overall sales.

Tractors producing in excess of 110 horsepower were appearing on the market and Case needed a machine to compete. The result was the 1200 Traction King, a four-wheel-drive, four-wheel-steer behemoth weighing 16,500 pounds. Its 451ci turbocharged engine produced 120 horsepower. The 1200 reestablished Case as the pre-eminent builder of big wheatland tractors and allowed the struggling company to enter the 1970s with a strong position in the marketplace.

ABOVE: Both axles were steerable on the 2670. The rear axles could be turned and locked into position, allowing it to crab across hillsides.

ABOVE RIGHT: After the 1985 acquisition of IH, Case tractors were painted red and adopted the name Case-International.

The 1970s In 1970 Case became a wholly owned subsidiary of Tenneco Inc. of Houston, Texas. With sales up and financial pressures eased, the next decade looked to be a good one. The profitability of small tractors was limited, and so Case positioned itself as the big tractor specialist. The 1200 had been given a cab and a horsepower boost to 144 in 1969, but was

introduced as the 1470 in 1970. The rest of the line was reintroduced as the –70 series, with cabs with rollover protection and operator comfort high on the list of new features.

The first truly new tractor of the 1970s was the 1270. With the majority of new tractors being ordered with cabs, it was time to build a tractor from the ground up with the cab in mind. The 1270 was designed so the cab would not have to be removed for any but the most serious repair procedures. At 127 horsepower, the 1270 was the largest two-wheel-drive tractor Case had ever built, but not for long. The 1370 soon followed with 143 horsepower, but it was surpassed in 1975 by the 180-horsepower 1570.

A 122hp two-wheel-drive tractor, the 1170, was added to the line in 1970. The 1170 and 1270 were given lavish introductions with specially painted demonstrator models. These black demonstrators were standard production models that had been carefully tuned and weighted for peak performance. A mystique developed around the tractors, with many insisting that they had specially built, hot-rodded engines. The black demonstrators became highly sought after and even today command a premium price. A 174hp four-wheel-drive tractor topped the tractor list in 1972. By 1978, Case's biggest tractor, the 2870, topped 250 PTO horsepower.

Case's concentration on large tractors was made possible by the purchase of David Brown Tractors of Meltham, England. Production of the medium-size 470 and 570 tractors was terminated, and these fine imported tractors served the 43- to 65-horsepower market. Once again Case consolidated the look of its tractor line in 1974 by combining David Brown's white tin with Case's Power Red chassis color.

Models continued to proliferate as Case refined and developed its line to meet the needs of American farmers. Nineteen seventy-eight was a year of consolidation. The six big row-crop tractors were replaced with four models of 108 to 180 horsepower. Mechanical four-wheel drive was offered as an option on these row-crop tractors for the first time.

over 50 models, which had to be pared down to a dozen or so. On the large tractor end of the scale there wasn't much question of which models would be kept. The four big Case-designed four-wheel-drives joined IH row-crop and utility tractors built in Europe to make up the Case-IH line.

In 1987 Case acquired the Steiger company, an American builder of big four-wheel-drive tractors. Steiger had a successful line, and Case-IH continued it. Mid-size tractors were modernized in 1988 when the Magnum series of 130hp to 195hp tractors was introduced. The Magnums were a combination of CDC engines and IH transmissions.

The 1990s Recognizing the need for a world tractor, Case-IH now introduced the Maxxum series. With Case transmissions and CDC engines, the Maxxums were flexible enough to incorporate the different braking and safety features required of tractors used in Europe.

In May 1999 plans were announced to merge Case with what was left of the Ford tractor concern, New Holland. New Holland was owned by the Italian company Fiat-Geotech, making Case a primarily foreign company. The merger was consummated in November, 1999. Case missed being the only American company to begin and end the century as a builder of tractors for the American farmer by one month.

The 1980s Case had essentially a new tractor line in place at the start of the decade. Five models from David Brown filled the lower horsepower end of the spectrum. Full-time four-wheel-drive models joined the two-wheel-drive row-crop tractors in the 100-130 horsepower range and three models of four-wheel-drive four-wheel-steer tractors of up to 250 horsepower.

The dismal farm economy of the early 1980s had been brutal to the farm tractor industry. One casualty was the venerable International Harvester Company. Case acquired parts of the tractor business from IH in 1985. With both the IH and Case lines of tractors, Case-IH had

RIGHT: After purchasing the company, Case upheld the Steiger tradition of building huge tractors with machines such as this 375hp 9280.

CHAPTER THREE

THE AMERICAN TRACTOR

★

FERGUSON

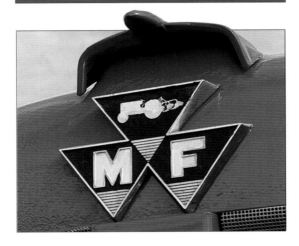

ABOVE: This simple logo has some of the biggest names in farm history behind it, Harry Ferguson, Daniel Massey, Alanson Harris, and Henry Ford, as well as hundreds of brilliant support engineers.

Harry Ferguson was well known throughout North America long before his first tractor was manufactured here. Ferguson had brought his three-point hitch with hydraulic draft control to Michigan in 1938 and demonstrated it for Henry Ford. Ford was impressed, and the two made an arrangement whereby Ford would manufacture a newly designed tractor incorporating the Ferguson hitch system. The tractor was called the "Ford Tractor-Ferguson System Model 9N," but was more widely known as the "Ford 9N."

The agreement between Ford and Ferguson allowed for Ferguson to design a line of special implements to go with the tractors and distribute both the tractor and the implements to dealers. It lasted until 1947, when

new management at Ford decided the tractor effort would be more profitable without Harry Ferguson's involvement. The agreement between Ford and Ferguson was terminated in November 1946, though Ford was to provide tractors for Ferguson through June 1947. Ford redesigned the tractor, designed its own line of implements, and began marketing them through its own Ford tractor dealers.

Standard Motors had been building a Ferguson tractor in the United Kingdom for Harry Ferguson since 1946. The Ferguson TE-20 had a 4-speed transmission and somewhat different styling, but otherwise differed little from the tractor Ford had been building. Ferguson quickly arranged to have 25,000 of them shipped to his dealers in the United States and began looking for someone to manufacture it in America.

With time ticking by and no manufacturer willing to take on the tractor, Ferguson laid plans to build the tractor himself in Detroit. The first Ferguson tractor manufactured in North America, the Model TO-20, came off the line on October 11, 1948. The

ABOVE: America's first introduction to Harry Ferguson was through the plow he designed for the Fordson tractor. This demonstration took place in Norfolk.

LEFT: In their famous meeting at the Ford farm in 1938, Harry Ferguson (left) used models to demonstrate his three-point hitch to Henry Ford (right).

"The blunt truth about this relationship is that it made Mr. Ferguson a multi-millionaire and cost the Ford Motor Company $25,000,000 in the process." Henry Ford II

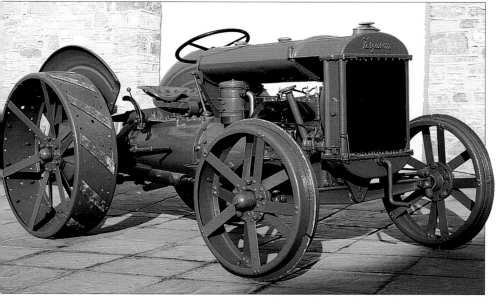

LEFT: *The Ferguson-Brown tractor embodied many of the features of the Ford-Ferguson and the famous Ferguson TO-20 which followed.*

ABOVE: *Without Ford's involvement in the enterprise, Ferguson was free to make his 20 series tractors exactly as he wanted them.*

TO-20 used a Continental Z120 engine. It weighed about 2400 pounds and had a 4-speed transmission. It was a direct competitor for Ford's 8N, and with its more modern and efficient overhead valve engine, was a slightly superior machine. Unfortunately, material restrictions due to the Korean War kept the Ferguson from realizing its sales potential.

The 1950s In the Fall of 1951 a slightly larger model, the TO-30 was introduced. The principal change was a larger 129ci Continental Z129 engine giving 29 horsepower, and a more robust drivetrain. Ferguson also had the Model TO-35 under development. Introduced in 1955, the TO-35

In December 1957 the lines were merged into what would, for a time, be the world's dominant tractor manufacturer—Massey-Ferguson.

BELOW: Harry Ferguson set out to make his 20 series world tractors. Though the TO-20 was the American version and the TE-20 was one of many British variants, this TO-20 displays UK registration.

ABOVE: It never attained the popularity of its rival Ford 8N, but with its overhead valve engine, the TO-20 was a somewhat more advanced tractor. Even children loved it (right).

followed the same design philosophy of all Ferguson tractors. It was low to the ground and lightweight, and it had the famous Ferguson hydraulic lift system. It had a Continental K134 engine producing 34 horsepower and a dual range, 6-speed transmission. A Deluxe model was available that had a tachometer and live power take-off. Both the standard and Deluxe had position control of the lift arms as well as draft control.

While the new tractors were being developed on the drafting tables, big changes were being made in the board room. Ferguson had always sought contract manufacturers for his tractors, and only owned his own plant in the U.S. because no manufacturer could be found quickly enough. One of the firms Harry Ferguson had contacted to build his tractor was the Massey-Harris Company. Massey-Harris had rejected Ferguson's proposal

RIGHT: Ferguson had a full line of three-point implements designed when the TO-20 went into production. Some took full advantage of the three-point hitch, and some would have been better left as standard pull-type implements. Ferguson is said to have made more money selling the implements than from selling tractors.

as "too risky," but subsequent developments revealed the risk was in not doing business with Ferguson. Ferguson's tractor was superior to any model in Massey-Harris' small tractor line, and advanced new models to compete with its larger tractors were on the drawing board. When Harry Ferguson approached James Duncan, president of Massey-Harris, with a plan to sell his North American business, the Massey-Harris board of directors countered with a plan that also included Ferguson's considerable worldwide interests. Ferguson found the plan agreeable, and on August 4, 1953 merged his company with Massey-Harris, creating Massey-Harris-Ferguson Ltd.

With this merger, which was in fact more of an acquisition by Massey-Harris, Ferguson had a solid manufacturer for his tractors and Massey-Harris had the world-famous three-point hitch. Massey-Harris-Ferguson adopted a policy of maintaining two separate lines that incorporated common tractors. Massey-Harris Ponys were being painted gray and sold by Ferguson dealers. A stretched version of the new Ferguson TO-35 was painted red and sold as the Massey-Harris 50. The 50 was then painted cream and gray and sold by Ferguson dealers as the Model 40. This confused the market (no-one knew what they were buying from whom), increased costs, and hurt sales. In December 1957 the lines were

merged into what would, for a time, be the world's dominant tractor manufacturer—Massey-Ferguson.

The corporate color for Massey-Ferguson tractors was changed to red with a metallic gray chassis. Henceforth all Massey-Ferguson corporation tractors could be easily identified by color. The new company then embarked on an aggressive campaign to provide tractors for every segment of the market. With the Model TO-30 (a Ferguson TO-30 painted Massey-Ferguson colors), and the Model 50, the small tractor end of the market was covered. For the mid-sized tractor market Massey-Ferguson developed the Model 65. In general design the 65 was similar to the 30, with a 50-horsepower Continental engine and a 6-speed transmission. The 65 introduced Massey-Ferguson's first entry in the shift-on-the-go transmission field. The Multi-Power option gave 12 transmission speeds in two ranges and shifting between ranges could be done without stopping the tractor. This feature was also made available on the 35.

For big tractors the company went outside to Oliver Corporation and Minneapolis-Moline. From Minneapolis-Moline they purchased the Gvi, added distinctive corporate paint and sheet metal, and marketed it as the Massey-Ferguson Model MF95. This brute of a tractor had a six-cylinder engine of 425ci displacement and weighed over 8000 pounds. In 1959 and

"The MF1150—World's first V-8 Row Crop Tractor"

– Promotional slogan on a 1960s dealer sign

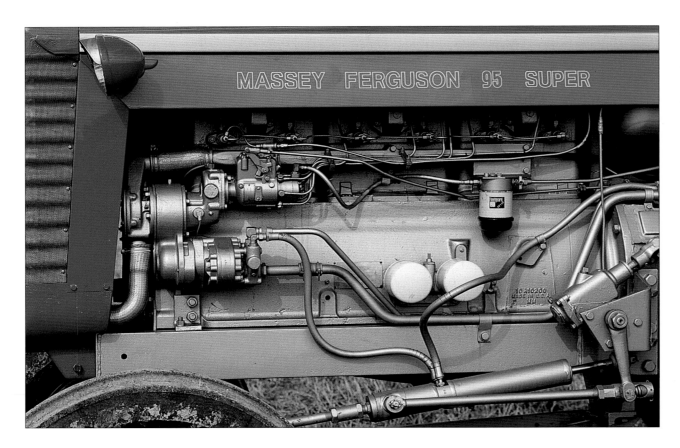

BELOW: The arrangement with the 95 Super was so successful that Massey-Ferguson began purchasing the 705/706 model the following year for sale in the U.S.

ABOVE: The Massey-Ferguson 95 Super was sold only in Canada and overseas. It was built by Minneapolis-Moline, and was no more than a Gvi repackaged in Massey-Ferguson sheet metal and paint.

RIGHT: After loading the spreader with the hydraulic loader, the farmer could pick up the spreader tongue with the hydraulic hitch and take the spreader to the field. Ferguson hydraulics had turned a dreaded, back-breaking job into one that could be done without leaving the tractor seat.

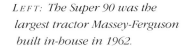

SUPER 90 BIG

Ferguson System tractor for 'broad-acre' farming

LEFT: The Super 90 was the largest tractor Massey-Ferguson built in-house in 1962.

1960 they bought 500 of the Model 990 from Oliver and marketed it as the MF98; these gave Massey-Ferguson 60-horsepower tractors for western wheatlands

The company did not like buying tractors from other manufacturers, and set about designing its own machine for the big tractor market. By 1959 it was ready. With a 60-horsepower gas or diesel four-cylinder engine and an 8-speed transmission, the new tractor gave Massey-Ferguson a big wheatland machine as the Model MF88 and also provided an entry into the growing 4-5 plow class of row-crop tractors with the MF85. The MF85 could be equipped with an adjustable wide front, a tricycle front end.

The 1960s Still not content with its position at the high end of the horsepower scale, Massey-Ferguson then decided to purchase 100-horsepower tractors from Minneapolis-Moline. These were sold as the MF97, and came in two- and four-wheel-drive versions. One new model, the 68-horsepower MF90 was introduced in 1962 to fill out the line while an entire new line of tractors was developed.

In 1964 and 1965 the new line of "Red Giants" tractors was introduced. Know internally as the DX project, the new range included the MF135 and MF150 models with 37 horsepower, the MF165 with 50 horsepower, the MF175 with 60 horsepower, the MF1100 with 90 horsepower, and the MF1130 with 120 horsepower. The DX tractors were designed for international markets, with the small and medium models being built in Europe and the two big tractors built in North America. The new tractors incorporated an Advanced Ferguson System with more sophisticated control and capacity for ever larger implements. Operator comfort and stylish looks were high on the list of features.

Multi-Power transmissions became standard on most models. The MF135 Standard replaced the venerable MF35, while the MF135 Deluxe added more horsepower from a 145ci Continental engine to the same chassis. The new MF165 retained the Continental 176ci gas engine when it replaced the MF65 in the line-up. The American engines were replaced by Perkins three-cylinder gas and diesel units of similar power in 1969, ten years after Massey-Ferguson purchased Perkins. MF175 and MF180 models filled the five-plow tractor category. Using identical engines and transmissions, the MF175 was a low-platform tractor in the old Ferguson style, while the MF180 was a traditional, high-platform row-crop tractor. Both were available with 62-horsepower gas or 63-horsepower diesel engines. A gap in the sub-25-horsepower area was filled in 1966 by the MF130, imported from France.

The horsepower race was spiraling upward and competition in that area could not be ignored. The demand for ever larger tractors by

ABOVE: The Model 399 was built in Massey-Ferguson's Coventry, England factory in 1987 through 1998.

American farmers was addressed again in 1965 when the first turbocharged Massey-Ferguson engine appeared in the 120-horsepower MF1130. Three years later the MF1150 appeared with a 135-horsepower turbocharged V-8 diesel.

New styling and quieter, more comfortable cabs came to the 80-horsepower and larger tractors in 1971.

Massey-Ferguson stepped up to four-wheel-drive in 1969 with the MF1500 and MF1800 tractors. With 153 and 171 horsepower respectively, these big articulated tractors were powered by Caterpillar V-8 diesel engines.

The 1970s New styling and quieter, more comfortable cabs came to the 80-horsepower and larger tractors in 1971. The improved models were distinguished by the addition of "5" to their model numbers. While all models retained the same basic engines, the MF1100 received a turbocharger and 10 extra horsepower when it became the MF1105. Both the MF1505 and MF1805 also received small power boosts.

ABOVE: The 130hp 8120 could be purchased with two-wheel-drive or, for an additional $10,400, with front-wheel-assist.

In 1973 the sub-100-horsepower tractors received the styling to match the larger tractors. Each tractor received a new model number and a horsepower increase. The 34-horsepower MF230 was the smallest and the MF285, with 82 horsepower, was the largest. In between were the MF235 (41 horsepower), MF255 (50 horsepower) and the MF265 (61 horsepower). In 1975 the MF235 was replaced by the MF245 of the same power and using the same engine.

With updated small tractors and big four-wheel-drive tractors in the fold, Massey-Ferguson engineers turned their attention to more powerful two-wheel-drive tractors. In 1975 the 160-horsepower MF2770 and the 190-horsepower MF2800 were launched. These tractors weren't on the market long before they were upgraded to 166 and 195 horsepower and

renamed the MF2775 and the MF2805. The mid-range two-wheel-drive line was upgraded in 1978 when the MF2675, MF2705, and the MF2745 replaced the MF1105, MF1135, and the MF1155. New styling and a small increase in horsepower were the major changes.

Competition in the four-wheel-drive market was strong, and a fierce horsepower war was being waged among the major manufacturers. In 1979 Allis-Chalmers, Case, John Deere, and Versatile all had four-wheel-drives that topped 200 horsepower. To be a player, Massey-Ferguson was going to have to do better than the MF2805. A new tractor series built around the Cummins 903ci V-8 engine was designed for 1980. These 30,000-pound tractors were virtually identical except for the state of tune of the engine. The MF4800 gave 179 horsepower, the MF4840 produced 211 horsepower, and the top of the line MF4880 put out an impressive 273 horsepower. For a short while, Massey-Ferguson was the horsepower leader in the North American market. Not content to rest on its laurels, the company produced the 324-horsepower MF4900 in 1981. These tractors came standard with a 12-speed transmission, but a 6-speed gearbox with three power-shifted ranges was available that gave 18 speeds.

BELOW: The British-built 3080 was available from 1987 through 1994.

The 1980s The tractor sales slump of the early 1980s hit Massey-Ferguson hard. The Detroit factory where the mid-range Massey-Ferguson tractors were built was closed in 1982. All Massey-Ferguson tractors except the four-wheel-drive models were imported from Europe. Mid-size tractors came from France, and with them came a feature that was unavailable on American-built tractors, front-wheel-assist. Three models were available: the 92hp MF3505, the 108hp MF3525, and the 127hp MF3545. A range of tractors with integral cabs was also available. These, the MF670, MF690, and MF698 shared mechanical components with the MF3505–MF3545 cabless tractors.

Small Massey-Ferguson tractors were sourced from England beginning in 1982. Model numbers changed, but the basic tractors remained the same. The MF240 replaced the MF230, the MF250 replaced the MF235, the MF270 was an MF255 but the engine was retuned to give five more horsepower. The MF290 replaced the MF265 and the MF298 replaced both the MF275 and MF285. Massey-Ferguson no longer had a traditional high-platform row-crop tractor in the line. In 1985 Massey-Ferguson changed its name to the Verity Corporation to reflect its greater interest outside the farm tractor area. The name Massey-Ferguson was retained for agricultural

ABOVE: By the 1990s the influence of European imports on U.S. design resulted in the availability of four-wheel-drive on every size of tractor.

RIGHT: The 2680 was only available as a four-wheel-drive tractor. Note the spin-out wheels front and rear.

BELOW: The 9240 was built in the United States for sale in Europe.

products, but the shift in focus for the parent company did not bode well for the future of Massey-Ferguson.

The 50- to 100-horsepower tractors in the 200 series were scrapped in 1986 and replaced with the MF300 series of tractors. The MF300s consisted of five tractors from 49 to 90 horsepower. These tractors were built in England, while a parallel line of similarly sized tractors with front-wheel-assist was sourced from Landini in Italy.

Also in 1986 the MF600 series of integral cab tractors was replaced with the MF3000 line. These French-built tractors of 60 to 115 horsepower featured the state of the art Autotronic electronic control. The Autotronic system disengages the front-wheel-assist when not needed, and inappropriate gear selections by the operator are blocked. The PTO is also automatically disengaged when overloaded or when the engine is being started.

The early 1980s introduced a period of product shuffling in the Massey-Ferguson line. Massey-Ferguson was truly a global corporation with distributors throughout the world, and manufacturing plants on every continent. Massey-Ferguson imported small tractor models from all over the world, some for less than a year. This may have been financially advantageous to the company in the short term, but it destroyed product continuity for dealers. The plethora of different models of similar size and capability, with different service and parts requirements, made life difficult for dealers and customers and did nothing to improve sales through the later 1980s.

The 1990s In 1991 AGCO Corporation of Duluth, Georgia purchased the North American distribution and parts business from Verity. In 1994 the remaining assets of Massey-Ferguson's worldwide business were

Consolidation of the line and advanced new tractor designs put Massey-Ferguson in a strong position at the end the century.

sold to AGCO. AGCO set out to rationalize the Massey-Ferguson tractor line in 1995. All models from 85 to 180 horsepower were replaced with two series of tractors. The MF6100 series consisted of three tractors, from 86 to 111 horsepower. The standard transmission was a 16-forward and 16-reverse Speedshift, while the 32-speed Dynashift transmission was a popular option. Autotronic electronic systems control was standard equipment. The MF8100 series filled the upper end of the power range with three models ranging from 130 to 180 horsepower.

The 300+ horsepower articulated tractors were no longer offered. The market for these tractors was small, and was adequately covered by other AGCO makes. Emphasis was placed on smaller front-wheel-assist tractors that had worldwide market appeal. In fact, in a reversal of the trend established in the 1980s, the MF9240 was a front-wheel-assist, 215-horsepower tractor that was built in the United States and exported to Europe.

The smaller tractor line that had been filled with the 300 series tractors since 1986 was revisited in 1997. A radical new line of world tractors was designed to fill the popular under-100 horsepower tractor sizes. Gone was the tin-box-on-wheels look that had characterized tractor design since the mid-1960s. The new line saw the return of the concern for operator visibility that had guided tractor design in the 1930s. The 4200 series included six models from 55 to 99 horsepower. Four transmissions were available, with on-the-go shifting, forward/reverse shuttle, or full synchronized shifting, depending on the model. Front-wheel-assist was optional on all models. The smallest MF4225 came with a low profile, high visibility cab only 96 inches high. The 65-horsepower MF4235 was available only in the boxy standard cab/hood configuration. The mid-range tractors, MF4240 (75 horsepower) and MF4250 (85 horsepower) series were available in standard and high visibility configurations. The top of the line MF4263 and MF4270, at 90 and 99 horsepower each, were available only in standard configuration.

Four tractors were added in 1998 to send Massey-Ferguson into the new millennium with a thoroughly modern line of farm tractors. The

ABOVE: The Model 3690 is powered by a 449ci Valmet engine producing 169 horsepower. The transmission is Massey-Ferguson's 32-speed Dynashift. Front-wheel-assist is optional.

MF8270 and MF8280 were 200- and 225-horsepower, four-wheel-drive tractors that comprised the top end of the Massey-Ferguson line. Both tractors had 18x6 powershift transmissions and roomy cabs with over 60 square feet of window area. The small end of the line got two new models also, the 49-horsepower MF2210 and the 58-horsepower MF2220. While packing nearly the same horsepower as the MF4225, these tractors were simpler, lighter, and more compact. Available in two-wheel-drive or four-wheel-drive, they had 12x12 shuttle transmissions, making them more useful for close-in work.

Consolidation of the line and advanced new tractor designs put Massey-Ferguson in a strong position at the end the century. With advances in electronic controls, front-wheel-assist and four-wheel-drive, and an emphasis on sophisticated design rather than brute horsepower, Massey-Ferguson leadership has positioned itself for success in the 21st century.

CHAPTER FOUR

THE AMERICAN TRACTOR

★

FORD

The first Ford production tractor, the Fordson Model F, appeared in 1917. Henry Ford had experimented with farm tractors for over a decade, but had never gotten one to operate to his satisfaction. After years of experimentation the tractor still wasn't quite up to his liking, but the prodding of his engineers and a desperate plea for tractors from the British government persuaded him to go ahead with it. The first production Fordsons were exported to Britain in 1917.

ABOVE: This cutaway drawing of the Fordson tractor illustrates the simplicity of the Fordson power train.

The Fordson was not manufactured for domestic consumption until 1918. When Ford went into mass production with the Fordson, he applied all he had learned from his years of building the Model T automobile. Designed from the beginning to be cheap to build and to sell in the millions, Ford was able to market it for an incredible $395 when comparable tractors were selling for three times that amount. While it lacked features such as a governor that other tractor designers considered essential, it was the least expensive horsepower on the market by far. It caused a sensation, and within four years it was the most popular tractor in America by a wide margin.

Ford manufactured the Fordson in Detroit until 1928, when he abruptly stopped production. Manufacturing space was needed for the new Model A cars and trucks, and Henry had lost interest in the Fordson. It had not been updated in ten years and farmers, as well as his own engineers, were clamoring for more modern features. Henry was more interested in building something radically new than in incrementally improving the Fordson. The Fordson was still popular in England, so he transferred manufacturing to a Ford plant in Cork, Ireland and turned his attention to experimenting with new tractor ideas.

LEFT: Henry Ford began experimenting with tractors early on. This picture of his "autoplow" was made in 1907.

ABOVE: The Fordson wasn't quite all Henry wanted, but the huge demand for tractors forced him to put it into production anyway.

★

ABOVE: The rear fenders were initially
designed to prevent the tractor from
flipping over backwards.

ABOVE: Gasoline was used to
start and warm up the engine.

LEFT: The Fordson evolved constantly. Notice
the difference between this 1922 model's
radiator and the 1917 model to its left.

1918
Fordson F

The Fordson Model F is perhaps the most influential tractor in history. The Fordson was the first tractor on tens of thousands of American farms. Because it sold for as little as $285, less than half the price of any comparable tractor, farmers were willing to gamble on it to try tractor farming. The Fordson introduced farmers to the potential of power farming, though it didn't always live up to that potential. It had several limitations, including a lack of a governor, a primitive ignition system, and an infamously stubborn clutch. Worst of all, it had a tendency to rear up in front and flip over on its back. A properly maintained and operated Fordson could return a good value on investment though, and farmers bought over half a million of them. After cutting their teeth on a Fordson, many farmers moved on to more sophisticated tractors.

LEFT: Fordson badge made in Detroit.

Many tales are told of bruises, broken arms—even death—caused by the Fordson's vicious crank. A Fordson in perfect condition started easily, but a poorly maintained tractor was difficult and dangerous to crank.

The overall length of the Fordson was determined by the width of a rail car. To minimize shipping costs, Ford wanted to be able to ship tractors side by side down the length of a car.

RIGHT: The seat over the rear axle could be a hot spot. The worm gear rear drive generated considerable heat.

Unit construction, where the engine
and transmission cases form the
tractor's frame, helped make
the Fordson quite rigid
and light.

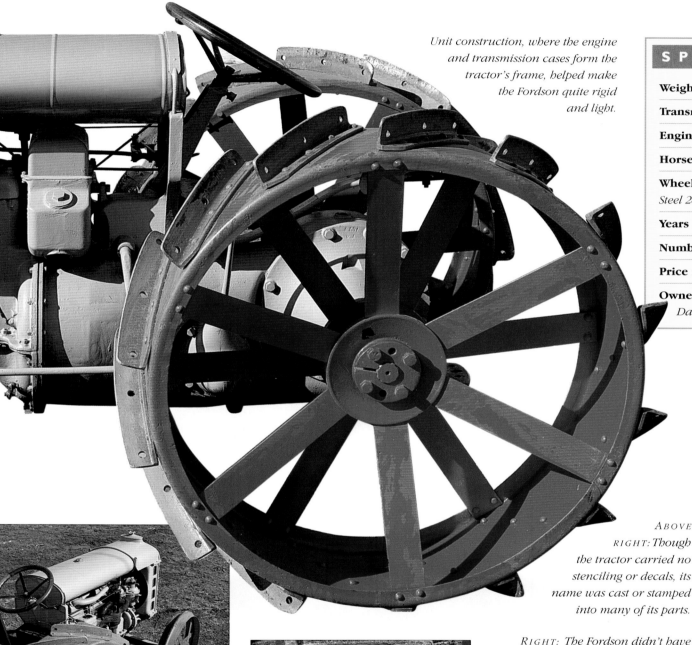

*ABOVE
RIGHT: Though
the tractor carried no
stenciling or decals, its
name was cast or stamped
into many of its parts.*

*RIGHT: The Fordson didn't have
a carburetor as such; this float
bowl is part of the mixer system
which regulated fuel flow.*

*LEFT: Gasoline was only used
for starting the engine. The
tractor ran well on kerosene
once the engine was warm.*

The 1930s With a water pump, high tension magneto, and a governor, the British-built Model N Fordson was a considerably better machine that the Model F. George and Eber Sherman, who had been importing Harry Ferguson's plows from Ireland for the Fordson for several years, soon began importing the Model N for the American market.

Row-crop tractors were in great demand in the United States in the 1930s. With most of the technical problems solved, the Shermans felt the

ABOVE: When production moved from Cork, Ireland to Dagenham, England the color of the tractors was changed to the traditional color of Essex farm carts—blue and orange.

only thing keeping Fordson sales down was its standard-tread configuration. In 1935, Ford engineer Howard Simpson set about designing a row-crop version of the Fordson. The result was the Fordson All-Around. Though the tractor resulted in some sales, it was crude and technologically out-of-date from the start.

In Ireland, Harry Ferguson had designed a tractor incorporating a revolutionary hydraulic three-point hitch. Eber Sherman, a friend of both Ford and Ferguson, arranged to have Ferguson demonstrate his tractor for Henry Ford. In October 1938 the Ferguson tractor was put through an impressive demonstration before Ford and his team of engineers. The Ferguson tractor was light in weight relative to its power, a factor that greatly impressed Ford. It was Ferguson's ingenious three-point hitch

LEFT: Harry Ferguson and Henry Ford, along with a group of engineers, pose with one of the first Ford-Ferguson 9N tractors. Ford and Ferguson were so keen to get the 9N project started, they failed to record clearly details of their agreement, an oversight both parties would later regret.

Henry Ford said, "Our competition is the horse." This tractor was meant to appeal to those farmers who were not mechanically inclined, found tractors noisy and difficult, and preferred to continue farming with horses.

with draft control that made it possible. Ferguson had concluded, rightly so, that a tractor did not need weight on the rear wheels to provide traction, what it needed was down force. His hitch allowed the plow to transfer some of the drag it exerted on the tractor to downforce on the drive wheels, providing traction without weight.

Ford had manufacturing facilities and know-how, and Ferguson had his hitch and a tractor design. The two quickly reached an agreement, which remained unrecorded, whereby Ford would manufacture a tractor incorporating Ferguson's hitch. Through his company, Harry Ferguson Inc., Ferguson would act as distributor for the tractor as well as design, have built by outside vendors, and distribute the specialized implements the tractor required. Teams of engineers from Ford and Ferguson set out to design a tractor, leaving the question of who was ultimately responsible for the tractor unanswered. After much discussion of whether it would be called a Ford, a Ferguson, or some combination, they settled on the cumbersome "Ford Tractor-Ferguson System." Most people just called it a Ford.

The new tractor was revolutionary in every way. The Model 9N, as it was called, weighed only 2340 pounds but with 13 drawbar horsepower it could pull a two-bottom plow in favorable conditions. The country's best-selling two-plow tractor at that time weighed 3700 pounds and had only one more horsepower. The 9N was specifically designed to be safe, quiet, and easy to

ABOVE: Henry Ford (center) made the landmark deal with Harry Ferguson. When his grandson Henry II (left) took control of the company, he ended the agreement.

operate. Henry Ford had once said, "Our competition is the horse." This tractor was meant to appeal to those farmers who were not mechanically inclined, found tractors too noisy and difficult to use, and who had preferred to continue farming with horses. To appeal to these farmers the

LEFT: The Fordson remained in production in England and was popular there, but it was so far behind American farm tractors that few were imported. This 1938 English Fordson was little improved over its 1929 counterpart.

1939
Ford-Ferguson 9N

Henry Ford had experimented with a number of unsuccessful tractor designs in the 1930s and found nothing to his liking until Harry Ferguson demonstrated his Ferguson tractor in 1938. Ford immediately saw its potential and entered an agreement to produce a similar tractor incorporating Ford's manufacturing know-how and Ferguson's hydraulic three-point hitch. The Ford Tractor with Ferguson System Model 9N, as it was called, was an immediate success. The 9N, with its low center of gravity, low operator station, adjustable wide front axle, and hydraulic three-point hitch, introduced a whole new tractor genre, the utility tractor. Within 15 years every major tractor manufacturer was producing a machine of the same general configuration. The 9N was only slightly changed to become the 2N in 1942, and was changed again to become the 8N in 1947. In all, over 750,000 were built, and it has been estimated that now, 60 years later, nearly half of them are still on the job.

LEFT: The first 700 to 800 9Ns were built with cast aluminum hoods. They were painted gray when new, but many restorers leave the aluminum unpainted and polish it.

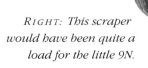

RIGHT: This scraper would have been quite a load for the little 9N.

The first 9Ns were delivered with Ford 8x32 tires, but later Ns came with 10x28s for more traction. By mounting the offset wheel disks and detachable rims in various positions, wheel spacing could be adjusted from 48 to 76 inches in four-inch steps.

SPECIFICATIONS

Weight (pounds): *2340*

Transmission: *3-speed*

Engine (displacement, rpm):
119.7ci, 2000

Horsepower (belt, drawbar): *20.29, 12.68*

Wheels and tires (front, rear):
4.00x9, 8.00x32

Years produced: *1939–1942*

Numbers built: *99,002*

Price new: *$585*

Owner of machine pictured:
Dwight Emstron

ABOVE: Ford's art deco styling was even extended to the radiator cap.

BELOW: Oil and air filters were standard equipment.

BELOW: Ford had its name on every part, even the battery and the tires.

LEFT: Before Dearborn took over the equipment business, Harry Ferguson sold a range of implements designed to take advantage of the three-point hitch.

9N was small and the driver sat down low to the ground, which must have given comfort to nervous first-time users. Starting the tractor didn't require running around the machine, cranking and fiddling with petcocks and spark adjustments. The operator just turned the key, pulled the choke, and pushed the starter button, all from the driver's seat. The little engine with its big muffler puttered quietly and was unlikely to scare horses—or farmers. The hydraulic draft control completely eliminated any chance of tipping over backwards, a serious hazard on most tractors.

Sales immediately took off. Within two years it was the best-selling tractor of its size in the world. Only one problem faced the venture. Though the price had risen to $685 Ford was still losing money on each machine he sold. The plan was to increase production to lower costs, but

World War II intervened, requiring a shift of focus to war production. Shortages of rubber and copper forced Ford to eliminate the battery ignition and starting, as well as the rubber tires. This required changes in the engine and chassis substantial enough to warrant changing the model name in 1942 to the 2N.

The Postwar 1940s The worldwide boom in tractor sales immediately after the war was particularly good to the 2N. Though the price had rocketed to over $1000, sales were double the nearest competitor, the Farmall A. Behind the strong sales figures, the picture

LEFT: When Ford took over distribution of its tractors, most dealers abandoned Ferguson.

BELOW: The NAA, introduced in 1953, was also known as the Golden Jubilee. It was a heavier, more powerful tractor than the 8N it replaced. With its obv engine it was also more fuel efficient.

wasn't nearly so bright. The only thing holding the operation together was the unwritten agreement between Henry Ford and Harry Ferguson, and Ford was dying.

Henry Ford had passed the leadership of the company on to his grandson Henry Ford II in 1945. By 1946 the younger Ford had discovered the true cost of the N tractor project to Ford Motor Company. It had lost over 25 million dollars in six years. His response was to create a new company, Dearborn Motors, in November 1946. The new company took over tractor distribution and the implement business from Harry Ferguson and informed Ferguson that after July 1947 Ford would no longer supply tractors to Harry Ferguson Inc.

Ford introduced an improved 2N, the 8N, in July 1947. The new tractor had a 4-speed transmission, better brakes, and a position control on the hydraulic lift. Sales skyrocketed

ABOVE: Though it used the same engine, the 8N, introduced in 1947, added a 4-speed transmission, position control for the three-point hitch, and better brakes to the 9N design.

immediately, with over 103,000 tractors sold in 1948 alone. The nearest competitor sold one-tenth that number.

Though experiments had taken place on a larger tractor, Ford ended the 1940s the way it began, with only one tractor model in its line. Ford was selling more of this one model than many manufacturers were selling from their whole line, so there was no hurry to make changes. The decade ended with the company finally making a profit on its tractors, which added to its complacency.

The 1950s Ford took the opportunity of its 50th anniversary in 1953 to introduce its most radical change to the N to date, the NAA. Popularly known as the "Golden Jubilee," the NAA was powered by a 134ci overhead valve, four-cylinder engine, Ford's first overhead valve tractor engine. Power was up from the 8N's 19.5 to 27.5. The new tractor was four inches longer, but only slightly heavier. Live power take-off and live hydraulics were offered for the first time. New sheet metal with a jet motif popular in the early 1950s continued to carry the 8N's light gray over red color scheme.

"Power that Pur-r-r-s when the going gets TOUGH!"

– 1953 Ford advertisement for the Golden Jubilee model

Since 1939 Ford had offered the farmer only one, very versatile tractor model. Management at Dearborn Motors ended that in 1953. The Golden Jubilee was introduced with such fanfare, an English import that also arrived at Ford dealerships was barely noticed.

The Fordson tractor had remained immensely popular in Britain. But by the end of World War II even the British felt it needed to be modernized. An hydraulic lift system and a ring and pinion-type rear axle

LEFT: The Golden Jubilee of 1953 celebrated Ford's fiftieth year in business. The model was replaced by the 600 and 800 series tractors after only one year of production.

ABOVE: The basic Model 600 was little different from the NAA Golden Jubilee, though a much needed 5-speed transmission and live PTO options were available.

RIGHT: The British-built Dexta carried the Fordson name on its radiator, but little of the venerable old design was left by the time it was introduced in 1958.

were added, as well as a new front axle and new wheels. Changes were so substantial it could be considered either a revised Fordson or a new tractor with a Fordson engine and transmission. Ford called it the Fordson Major. At 4300 pounds and 27 horsepower it was heavier and somewhat more powerful than the NAA, but offered none of the new NAA's sophistication. It was available with a diesel engine and with a high

clearance row-crop chassis, and that may explain some of its appeal to Ford's marketing people. Diesel power was the coming thing in all sizes of tractors, and Ford had no high-clearance narrow-front tractor.

In 1954 the 600 and 800 series of tractors were introduced. The 600 series was powered by the same 134ci Red Tiger engine used in the NAA. The 640 had a 4-speed transmission, the 650 had a 5-speed transmission

"America's Largest 'Pick up and Go' Family"

— Ford advertisement for the 800 series, mid-1950s

BELOW: The 861 Powermaster was Ford's most powerful domestic model.

BELOW: Ford's 861 combined a four-cylinder gas, diesel, or LP engine, a 5-speed transmission, and live PTO to create a very versatile, popular tractor. It was also available with Select-O-Speed or a simple 4-speed transmission.

LEFT: Ford was very successful with offset tractors designed for vegetable and specialty crop farmers.

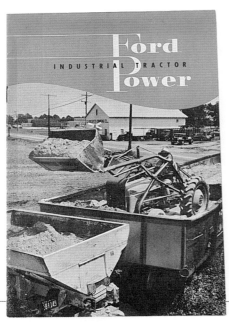

and standard PTO, and the 660 had the 5-speed and live PTO. The 800 series offered the same options behind a 172ci version of the same engine.

The N-type of tractor, with its high horsepower to weight ratio, three-point hitch, hydraulic system, and low operator position, came to be know as the "Utility" tractor. Ford had enjoyed a niche market with the N series of tractors, a niche that grew to comprise 16 per cent of the tractor market by 1951. Other manufacturers couldn't ignore a market that large, and began designing utility tractors of their own. In response, Ford, rather than defend its position as the premier builder of utility tractors, decided to meet them on their turf and expand into row-crop type tractors.

In the Fall of 1955 the two-plow 700 and three-plow 900 series of row-crop tractors were unveiled. These were essentially the 600 and 800 tractors reconfigured as row-crop machines, using the same engines and offering the same options.

In 1958 the Fordson Majors were replaced with a completely new tractor called the Power Major. A 220ci diesel engine provided 48 PTO horsepower. It had six speeds forward and two in reverse and could be ordered with live PTO. Ford imported another tractor from England in 1958, a modern small diesel called the Dexta. The Dexta was a utility type

ABOVE: The 6000 operator's
manual illustrates the
tractor's new comfort features.

ABOVE: Early 4000 series
tractors were updated versions of
the old 800 series in new sheet metal and
blue and gray paint.

tractor, just like the 600s, and at 31 PTO horsepower it was slightly more powerful than the 600/700 tractors. It was its three-cylinder Perkins diesel engine that made it unique. Ford made no attempt to integrate the British imports into its product line. They sat in the showrooms painted their native blue with orange wheels next to the gray and red Fords.

In 1959 Ford introduced its own four-cylinder diesel engines as well as LP fuel adaptations for the 600/700, and 800/900 series. A 500 series of

tractors used either the gasoline, LP, or diesel engine from the 600/700 tractors. These were low production specialty tractors with high clearance and chassis offset to the side to provide extra visibility for cultivating vegetable and specialty crops. With the introduction of the 500 series, Ford had gone in one decade from one model to a selection of four basic tractor sizes in over 50 chassis/transmission combinations.

BELOW RIGHT: The first Ford tractor to surpass 100 horsepower was the 1968 Model 8000.

Ford had earned a reputation for building small dependable tractors. That soon changed, and the change began in 1959, with the introduction of the Select-O-Speed transmission. Since the early years of the decade manufacturers had worked at developing transmissions that allowed shifting gears without stopping the tractor. Ford's Select-O-Speed was revolutionary because, for the first time, the operator could shift through the entire ten speeds without stopping the tractor. On paper, at least, Ford had a step on the competition.

Ford used the advanced transmission in its domestically produced tractors, where it suffered durability problems. While not a disaster, the Select-O-Speed was troublesome enough to earn Ford a reputation as a fragile tractor. This did not bode well for the company's plans to move into the big row-crop tractor market.

The 1960s In 1961 Ford consolidated its British and American operations and embarked on a unified international program of tractor design it called the "World Tractor." In North America this meant integrating the British imports with U.S.-produced tractors. Gray from the American tractors and blue from the British tractors were combined to form a

RIGHT: A buyer could have an 8- or 16-speed transmission for the 8000, but the troublesome Select-O-Speed was not an option.

corporate color identity, and styling was made similar enough that all the tractors could be identified as Ford products. The "thousand" model series was unveiled in the Fall of 1961 with the Model 6000 at the top of the line. The Fordson Super Major became the 5000, the old 800/900 series became the 4000, the 600/700 tractors became the 2000. The small domestically built diesel was dropped and the little Super Dexta became the 2000 Diesel. The Select-O-Speed transmission was an option on the smaller models

The 6000 was a medium-sized row-crop tractor using a 223ci engine of 66 horsepower or a 242ci diesel producing the same power. It was only available with the Select-O-Speed transmission, and it was a disaster. Recalls and field updates plagued the tractor from the beginning but it remained in production until 1967.

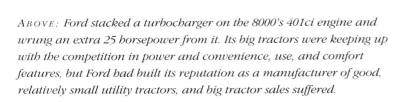

ABOVE: Ford stacked a turbocharger on the 8000's 401ci engine and wrung an extra 25 horsepower from it. Its big tractors were keeping up with the competition in power and convenience, use, and comfort features, but Ford had built its reputation as a manufacturer of good, relatively small utility tractors, and big tractor sales suffered.

*Ford capitalized on this reputation in the 1970s by building a
wide variety of special-purpose tractors while trying to keep up with
the trends in mainline agricultural and industrial tractors.*

*BELOW: In 1973 the 8000 received an update and a new name—the 8600. The
engine was retuned for more power and the operator's station was redesigned
to create the new model.*

In 1965 a major change took place in the smaller tractor line, though the
6000 remained little changed. The 5000 series became a domestically
produced tractor, with a 233ci four-cylinder gas or a domestically built 256ci
diesel. The 4000 series tractors were given a new 8-speed transmission and
a new engine, as well as being restyled in the contemporary squared look.
Its 172ci four-cylinder was replaced by a 192ci three-cylinder engine with six
more horsepower. A similar 3000 series of tractors with a 37hp three-
cylinder engine was also announced. The 2000 series was completely
redesigned and was available with either a 6- or 12-speed transmission and
a 158ci three-cylinder gas or diesel engine.

In 1968 the 8000, with 106 horsepower, was unveiled. The 8000 was a
wide-front row-crop tractor with a big, 401ci six-cylinder diesel engine and
choice of an 8-speed or 16-speed transmission. This was Ford's first attempt
to escape the legacy of the 6000. A more powerful version, the 131hp
turbocharged 9000, was also announced.

The 1970s The versatility of the 9N coupled with the familiar Ford
name had led aftermarket manufacturers to choose Fords as a platform
for a huge variety of specialty machines, from backhoes to golf course

grooming tractors. Ford capitalized on this reputation in the 1970s by
building a wide variety of special-purpose tractors while trying to keep up
with the trends in mainline agricultural and industrial tractors.

In 1973 Ford began a successful program of importing compact utility
tractors from Japan. The first was the Model 1000, a 23hp two-cylinder
diesel tractor weighing only 2300 pounds. This tractor was ideal for
landscapers, truck gardeners, and the growing number of hobby farmers
in the United States. Also in 1973 the 8000 and 9000 were given more
power and updated styling and renamed the 8600 and 9600.

In 1976 the Model 1000 was replaced by the Model 1600, essentially the
same tractor with different styling. Ford dropped the 2000 through 7000
models, replacing them with seven updated models. The 32hp 2600, the
40hp 3600, and the 52hp 4600 replaced the 2000, 3000, and 4000 series.
A 45hp 4100 and the 60hp 5600 were added to the line. The 5000 was
replaced by the 6600 at 70 horsepower, and the 84hp 7600 replaced the
7000. The two smaller tractors were available with either gasoline or diesel
engines. These tractors were available in a variety of chassis types, though
the most popular was what had come to be called an "all-purpose" style,
where the driver sat astride the transmission as in a utility tractor, but with
a little more ground clearance than a utility model.

Two new models were added in 1977, the 6700 and 7700. These were
Model 6600 and 7600 tractors with a redesigned operator's station that
placed the driver on a flat platform above the transmission. The 6600 and
7600 tractors continued. The 8700 and 9700 were 8600 and 9600s with the
redesigned operator's station were introduced the same year, but the old
style chassis was no longer offered.

Ford had concentrated on its World Tractor program, which had given
it a huge catalog of specialty tractors, but nothing approaching 200
horsepower. The market for large tractors was growing rapidly, but Ford
was in no position to capitalize on it. In 1978 Ford offered four articulated
four-wheel-drive tractors built by the Steiger Tractor Company of Fargo,
North Dakota. The four models, FW-20, FW-30, FW-40, and FW-60 used
Cummins V-8 engines and produced 210, 265, 295, and 335 horsepower
respectively. These tractors remained in the line until 1982.

In 1979 a new line of two-wheel-drive, flat platform tractors was
introduced. The TW-10 took the place of the 8700, the TW-20 took over
from the 9700, and the TW-30 introduced a larger size at 163
horsepower. Ford had developed a very successful business in the
compact tractor field and added five new models, the 11hp 1100, 14hp
1300, 17hp 1500, the 23hp 1700 (replacing the 1600), and the 27hp

RIGHT: Introduced in 1978, the 7810 used the same 401ci diesel engine first used in 1968 in the 8000. The engine was detuned to produce only 86 horsepower. Front-wheel-assist and a 16-speed transmission were optional.

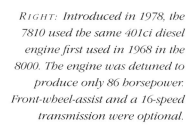

1900. The 1900 was powered by a three-cylinder diesel engine, the other four models were powered by two-cylinder diesels.

The 1980s Ford entered the new decade with a line of agricultural tractors that ranged from 11 to 335 horsepower. No changes took place in 1980, but 1981 saw a new 14hp compact tractor, the 1200 with four-wheel drive as standard, added to the line. The 5600, 6600, 7600, 6700, and 7700 models received increased engine displacement and power and were called the Blue Power Specials.

Bigger changes were in store for 1982. A new model, the 2310 was added. With 32 horsepower, this three-cylinder diesel filled the power gap between the 1900 and the 2600. Beginning with the 2600 and extending through the 7700, the tractor line received a small power boost and an optional 8-speed transmission. The largest three models also had the option of a 16-speed transmission. These tractors were identified by adding a "10" to their previous designation. The 5610 through 7710 also received new graphics with the Ford name displayed much less prominently—a portent of things to come. Ford ceased purchasing large articulated tractors from the struggling Steiger Tractor Company, leaving it without tractors for that market.

The TW series of tractors was revised in 1983. The TW-5, TW-15, and TW-25 took the place of the previous models. The TW-35, at 170hp, was a new size—Ford's largest two-wheel-drive tractor ever. All used the same 401ci diesel engine in normally aspirated, turbocharged (TW-15 and TW-25) and turbocharged and intercooled form. These tractors featured an 8-speed transmission and a 2-speed power shift, giving 16 speeds in all. Significant revisions were made at the small tractor end of the line. The compact tractor line was redesigned, with all but the 1100 (now called the

ABOVE: Equipped with optional front-wheel-assist and a cab, this TW-15 cost over $50,000 in 1987. It was powered by a turbocharged 121-horsepower version of the Ford 401ci diesel.

1110) getting new, three-cylinder engines. The new 1910 got a power boost to 29 horsepower. The 2610 was replaced with the 36hp 2910 while the 42hp 3910 replaced the 3610. Changes in the small tractors continued in 1984, with a new four-cylinder Japanese import, the 34hp model 2110 added to the compact line. The smallest domestically produced tractor, the 2310, was replaced by the 32hp 2810.

Ford Motor Company was clearly looking for a graceful way out of the tractor business.

LEFT: The 5610 was a popular utility tractor in the 70-horsepower range. It was available in two-wheel-drive, four-wheel-drive, and with or without a cab. The 5610 frequently saw use as a loader tractor.

BELOW: Although Ford Motor Company no longer had anything to do with the tractors, the Ford name was still used through 1999.

In 1985 Ford announced it was planning to transfer farm tractor manufacturing to Europe. Manufacture of three-cylinder tractors would be moved from Ford's Michigan plant to the Basildon, England facility while the TW series would be built at its Antwerp, Belgium plant. Ford also announced it was planning to buy New Holland, a short line manufacturer of primarily forage equipment, and change the corporate name to Ford-New Holland. Ford Motor Company was clearly looking for a graceful way out of the tractor business.

A new tractor using the 401ci diesel engine in normally aspirated form was added in 1986. This tractor, the 8210, was similar in type to the 7710, but had 100 horsepower.

In 1987 Ford-New Holland announced it was buying the Versatile Farm Equipment Company. The next year the Versatile purchase brought big articulated tractors back to the Ford-New Holland line. Four models, the

846, 876, 946, and 976, all previously Versatile models, were given the Ford Versatile name. They all used Cummins V-8 engines ranging in horsepower from 193 to 360. The models in the seven smallest 10-series tractors, the 1110 through the 2110, were replaced with the more powerful 20-series models. The 1120 through the 1520 came with two-wheel-drive and a 9-speed transmissions as standard. Front-wheel-assist and a hydrostatic transmission were optional. The three larger models came standard with two-wheel-drive and a 12-speed transmission. A synchronized transmission with direction reverser and front-wheel-assist were optional.

The 1990s Ford's mid-sized line of tractors was completely revised in 1990. A 32hp 3230 and 38hp 3430 used a 192ci three-cylinder diesel engine and 8-speed transmission. The 45hp 3930 and 55hp 4630 shared a 201ci three-cylinder diesel. The standard transmission was an 8-speed, with a two-speed power shift option that gave 16 total speeds. These tractors replaced the 2810 through 4610 models. Through Versatile, Ford-New Holland introduced a new concept in tractors to the American market. Versatile had developed a successful bi-directional tractor before its purchase by Ford-New Holland. With the addition of a Ford engine and other modifications it became the Ford Versatile 9030. The seat and control console could be rotated 90 or 180 degrees, so the tractor could be operated with equal ease in forward or reverse. The tractor had four-wheel-drive and three-point hitches at both ends to make maximum use of its 100 horsepower. Also new for 1990 were four tractors to replace the TW-5 through TW-35 series. The 8530, 8630, 8730 and 8830 had 105, 121, 140, and 170 horsepower respectively. The standard transmission was a 16-speed, but a programmable, electronically controlled, 18-speed full powershift unit was optional, as was front-wheel assist.

Few changes were made to the tractor line in 1991, as Ford Motor Company announced it was selling 80 per cent of Ford-New Holland to Fiat of Italy.

The 5610 through 8530 range was replaced in 1992 with the Powerstar series of tractors, though a low price 5610S with 66hp and a Genesis engine continued until 1996. This series introduced the all-new Genesis engines. The 5640, with the same horsepower as the 5610, offered many comfort and convenience options and a choice of an 8- or 16-speed four-range synchro transmission or a 12- or 16-speed powershift unit. A 76hp 6640 and an 86hp 7740 were available with the same features and options. The 90hp 7840, 96hp 8240, and 106hp 8340 used six-cylinder diesel engines. The standard transmission was a 12-speed, with a 16-speed four-

BELOW: *Henry Ford responded to the monster tractors of his day with the little Fordson, never dreaming his name would appear on a behemoth like this 22,000-pound 946.*

range powershift optional. All the new models were available with a cab and front-wheel assist.

1993 saw two tractors added to the compact line. The 22hp 1620 came with a hydrostatic transmission only. The 1715 used the same engine, but came with a 9-speed collar shift transmission. Front-wheel-assist was optional on both models.

Another sweeping change was in the offing for 1994. All four models at the big horsepower end of the line were replaced. The 9280, 9480, 9680, and 9880, had 250, 300, 350, and 400 engine horsepower respectively. A 12-speed transmission with four powershift speeds in each of three ranges was standard, though a 12-speed full powershift was available for the two smaller models. A whole new line of big two-wheel-drives was announced. The 8670, 8770, 8870, and 8970 shared the same 456ci turbocharged engine in states of tune that gave them 145, 160, 180, 210 horsepower. The standard transmission was a 16-speed, electronically controlled, full powershift. Front-wheel-assist was available, as was an axle that shifted to the outside to allow tighter turning. A 62hp utility tractor, the 5030 was added to the line, as was the 1215, a 13hp compact tractor.

In 1994 Ford Motor Company divested itself of its interest in the tractor division completely. The Ford name was dropped and the official company name was changed to New Holland. Though there was in effect no longer a Ford tractor, when Fiat-Geotech bought the remaining interest in Ford-New Holland, an agreement was reached whereby Fiat was allowed to use the Ford name on tractors through the year 2000. As the end of the century approached and new tractors were designed, the name Ford was used less and less. In May 1999 New Holland joined with Case to form the largest farm equipment manufacturer in the world with New Holland's parent company Fiat holding a 71 per cent interest.

Henry Ford's Fordson had introduced hundreds of thousands of farmers to mechanized farming. The Ford N tractors had brought new levels of comfort, safety, and ease of use to the farm tractor. In the year 2000 60-year-old Ford tractors were still at work on American farms. Ford, the name that earned such a prominent place in farm history, had disappeared from the showrooms, but Ford tractors in faded shades of red, gray, and blue still dotted the North American landscape as the century closed.

Chapter Five

The American Tractor

★

Hart-Parr / Oliver

ABOVE: Hart-Parr was among the first manufacturers to offer a reasonably dependable tractor, giving some legitimacy to the claim, "founders of the tractor industry."

One of the great names in farm tractor history got its start in 1896 when Charles Hart and Charles Parr, students at the University of Wisconsin, combined their efforts on a college thesis on internal combustion engines. The very next year the young Hart and Parr were manufacturing engines, and by 1901 had organized in Charles City, Iowa as the Hart-Parr Company. In 1902 they built their first gas traction engine.

Unlike the tentative experiments with the tractor business on the part of dozens of other manufacturers, Hart and Parr were in the business to stay. They left no doubt of their intentions when they placed their first advertisement in the December 1902 issue of *American Thresherman*. Farmers, threshermen, and steam engine builders alike could no longer disregard the coming of the gas traction engine. The determination of Hart and Parr, along with the superiority of their tractor drove company growth at a tremendous rate.

The First Tractors The first two machines, built in 1902, were experimental units. In 1903 their first production tractor was built. This machine weighed 15,000 pounds with a two-cylinder engine that displaced 2042 cubic inches. The flywheel alone weighed half a ton. It produced 18 drawbar horsepower and 30 belt horsepower. With hit-and-miss governing and make-and-break ignition,

ABOVE: The Little Devil was an unfortunate exception to Hart-Parr's generally sound offerings. Its failure did little to advance the fragile new industry.

BELOW AND RIGHT: Hart-Parr's advertising demonstrated the variety of work its tractors were capable of.

LEFT: The 18-36 was the ultimate development of the 15-30, which debuted in 1918. The 15-30 received frequent updates in its life, eventually, as the 18-30 I, the design was developed as far as was practical.

BELOW: Introduced in 1928, the 28-50 was the largest of the Hart-Parr crossmotors. Its engine was built from two 12-24 H blocks and displaced 674 cubic inches.

ABOVE: Hart-Parr was already at work on a new generation of tractors when this 12-24 was built in 1926.

the tractor was crude, but remarkably reliable. This tractor more than any other contributed to the early success of the gas traction engine.

Fifteen of the 17-30 tractor were built before a run of 22-45 tractors based on a second design was manufactured. By 1907 400 gas traction engines had been built. 1907 was a watershed year. The 17-30 had been developed into the 30-60 and orders were pouring in for the big tractor. And it was now a "tractor." Hart-Parr had begun using that term in their advertising instead of the cumbersome "gas traction engine."

For the next few years Hart-Parr experimented with a number of smaller tractors, some more successful than others. A 15-30, that was later re-rated to 20-40, was one of the successes. The Oil King, an 11,500-pound kerosene-burning tractor of 18-35 horsepower, was a successful mid-sized tractor. The Little Devil, Hart-Parr's smallest tractor, was a notable failure. This unique tractor had a two-cylinder, two-cycle engine of 333 cubic inches. It weighed just over 6000 pounds—light for a Hart-Parr. The transmission had two speeds that turned the single drive wheel in opposite directions. To change transmission speeds the operator shifted

gears and changed the direction of rotation of the engine. With a two cycle this was relatively easy, and this gave two forward and two reverse speeds. These tractors were notoriously unreliable, and Hart-Parr spent an untold amount of money recalling the machines.

While the company was publicly struggling with the embarrassment of the Little Devil, behind the scenes it was developing a progressive new line of tractors that would carry the company for over a decade. The new tractors were introduced in 1918 and were an immediate hit. The first new model to be unveiled, the 12-25, was quickly rerated to 15-30 horsepower. It weighed only 5450 pounds but could do the work of a machine weighing twice as much. It had a two-cylinder throttle-governed engine of

1913
Hart-Parr 30-60

In the infancy of the tractor industry, farm tractors were notoriously unreliable. The market was rife with experiments masquerading as field-ready tractors. When the inevitable breakdown occurred and the hapless owner turned to them for support, he would often as not find the manufacturer had gone out of business and fled the territory. The Hart-Parr 30-60 changed all that. The 30-60, introduced in 1903 as the 22-45 and uprated to 30-60 in 1907, developed a reputation as a reliable machine backed by a dependable company. A Hart-Parr built in 1903, a period when a tractor's life was often measured in months, was still in daily operation 21 years later when the company bought it back from its owner for promotional purposes. Hart-Parr experimented with many unsuccessful designs throughout the 1910s, and it was the stamina and reputation of the 30-60 that carried the company until the 1918 introduction of its very successful lightweight tractors.

BELOW: Exposed gearing was a weakness of all early tractors. Even the steering gear wore quickly.

LEFT: Even tractor manufacturers had a sense of style, dressing their tractors up with pin striping where possible.

RIGHT: Early farm tractors were expected to spend a large part of their career doing belt work. As its massive belt pulley attests, the 30-60 was no exception to this rule.

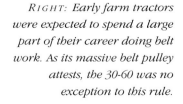

The operator's station did not include a seat. He was expected to be too busy with the controls to spend much time sitting.

SPECIFICATIONS

Weight (pounds): *18,000*

Transmission: *1-speed*

Engine (displacement, rpm): *2356ci, 300*

Horsepower (belt, drawbar): *60, 30*

Wheels and tires (front, rear): *c.66in*

Years produced: *1907–1918*

Numbers built: *3445*

Price new: *$2600*

Owner of machine pictured:
Gary Spiznogle

The 30-60 earned the name "Old Reliable" by being one of the first gas traction engines that could be depended on to start and run on demand.

RIGHT: The Hart-Parr 30-60 was started on gasoline and operated on kerosene after the engine warmed up.

The 30-60 is a giant – twice the height of a man and weighing nine tons.

RIGHT: Charles Hart and Charles Parr studied together at the University of Wisconsin and later established the company that bore their names.

RIGHT: *Toy models of the 70 are still popular.*

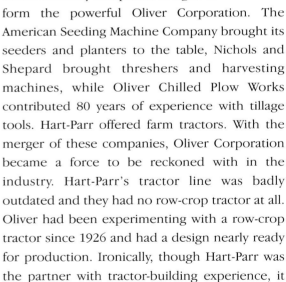

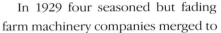

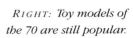

only 465 cubic inches. The 15-30 (also known as the 30) was joined in 1920 by a 10-20 hp tractor known as the 20. In 1923 a larger tractor, the 40 (22/40 horsepower) joined the line. It was basically a larger tractor with two Model 20 engines. A much improved version of the 15-30 was introduced in 1924. This tractor had fully enclosed gearing and was rated at 16/30 horsepower. These three tractors, the 12-24, 16-30, and 22-40, made up the Hart-Parr line until the 28-50 was introduced in 1928. The 16-30 was the most popular, though Hart-Parr sold only 3149 of this model, while Rumely sold 4855 of its similar OilPull Model L. With periodic updates that included horsepower increases and a change to 3-speed transmissions, this line of tractors was continued until 1930.

In 1929 four seasoned but fading farm machinery companies merged to form the powerful Oliver Corporation. The American Seeding Machine Company brought its seeders and planters to the table, Nichols and Shepard brought threshers and harvesting machines, while Oliver Chilled Plow Works contributed 80 years of experience with tillage tools. Hart-Parr offered farm tractors. With the merger of these companies, Oliver Corporation became a force to be reckoned with in the industry. Hart-Parr's tractor line was badly outdated and they had no row-crop tractor at all. Oliver had been experimenting with a row-crop tractor since 1926 and had a design nearly ready for production. Ironically, though Hart-Parr was the partner with tractor-building experience, it was an Oliver-designed tractor that first went into production. Hart-Parr's contribution was its manufacturing expertise and facilities.

The 1930s The first Oliver tractor to reach dealers was the Oliver-Hart-Parr Row-Crop. This was a row-crop tractor of 18/27 horsepower. It had a single front wheel and rear wheels that were adjustable by sliding them in and out on a splined axle. This rear axle design would eventually become

ABOVE: *Oliver-Hart-Parr's only row-crop tractor until 1935 was the 18-27. This 1932 model is the improved version, with dual front wheels and improved steering. Oliver experimented with a diesel engine in this model in the late 1930s, but never put it into production.*

ABOVE TOP: *The largest of the new tractors was the 28-44. This tractor was much more compact and easy to operate than the crossmotor 28-50 it replaced. It was a favorite for belt work and an excellent plow tractor as well.*

> *"The Oliver Standard 70. It has everything—*
> *it does everything—it's a beauty."*
>
> *– From the sales brochure for the Oliver Standard 70*

LEFT: The Oliver 70 came in standard-tread and row-crop versions, as well as an industrial version and a special aircraft tug called the Airport 25.

Model B became the 18-28. Both of these were offered in a variety of configurations and were popular industrial tractors. In October 1931 a Row-Crop with dual front wheels and improved steering and brakes replaced the original Row-Crop. It was about this time that it began to be called the 18-27.

In 1935 Oliver introduced the Model 70. This was an advanced row-crop tractor, with a high compression six-cylinder engine, and a 4-speed transmission. Oliver, in conjunction with the Waukesha Motor Company, which was responsible for Oliver's engines, introduced the first high compression gasoline engine in a farm tractor. The engine was available in low compression form for burning distillate. A standard-tread version of the 70 soon followed.

1937 was a watershed year for Oliver tractors. Since the merger of the companies in 1929 the name appearing on the tractors had been Oliver-Hart-Parr. In 1935 the Hart-Parr names began to be phased out. By 1937 all tractors were known simply as Oliver. The 70, already a stylish tractor, received even more streamlined sheet metal. The original Row-Crop tractor and the 18-28 were given the common name, the Model 80, and the 28-44 was renamed the Model 90. A Model 99, based on the 90 but with a high compression gasoline engine, was added to the line.

ABOVE AND RIGHT: Soon after the 70 was introduced, the 18-27 became the 80 Row-Crop and the 18-28 was renamed the 80 Standard. A Standard version of the 70 was also available.

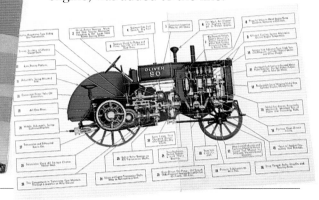

the standard for row-crop tractors. Shortly after the unveiling of the Row-Crop, the Model A 28/44hp standard-tread tractor was introduced. An 18/28hp standard-tread tractor based on the Row-Crop's engine was soon added to the line. The Model A soon became known as the 28-44 and the

1943
Oliver 70

The Model 70 used a innovative, high compression six-cylinder engine. Its six cylinders running at high speed (for the time) gave the 70 an unprecedented smoothness. The higher compression allowed the engine to develop power more efficiently, but it also required high octane gas. High octane fuel was more expensive, so the economic benefits of higher efficiency were little realized. The engine was also available in a very low compression distillate version. Before the 70 was introduced, several tractors painted a variety of colors, red, green, orange, silver, and gold, were sent to fairs around to country to gauge public reaction. The green tractor was the most popular, so the tractor was painted green with an orange stripe. The Oliver-Hart-Parr 70 was the last new tractor to carry the Hart-Parr name.

Standard Oliver steel wheels were just a skeleton framework that the lugs attached to. Oliver called them "Tip Toe" wheels because the tractor rode on just the tips of the lugs on firm surfaces. This tractor has road bands to smooth out travel on hard roads.

RIGHT: The Hart-Parr name began to disappear from Oliver tractors in the mid-1930s.

SPECIFICATIONS

Weight (pounds): *3340*

Transmission: *4-speed*

Engine (displacement, rpm): *201ci, 1500*

Horsepower (belt, drawbar): *24.95, 15.89*

Wheels and tires (front, rear):
27x4.40, 59.50x5.625

Years produced: *1935–1948*

Numbers built: *65,000*

Price new: *$915*

Owner of machine pictured:
Brian Brown

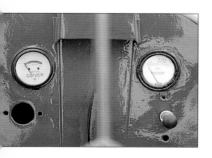

LEFT: The dash-board has only water temperature and oil pressure gauges and a choke knob. On electric start models, a starter button occupies the vacant hole on the left.

The fenders, road bands on the rear wheels, and the wide front axle were all optional equipment that pushed the new price of this Oliver 70 to over $1000. This tractor does not have an electric starter, which was a popular option.

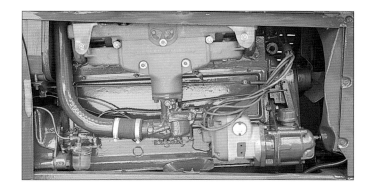

ABOVE: This powerplant was one of the first high compression engines used in a farm tractor.

RIGHT: Forward visibility from the operator's seat was excellent, an important consideration when cultivating row crops.

BELOW: Oliver's mounted cultivators were a simple design that fit on pipes that passed through round holes in the frame.

"We produced the first successful oil tractor and have been building tractors continuously ever since. We are specialists. We have learned one thing and learned it well." – Advertisement for the Hart-Parr 30

The 1940s Oliver began the new decade with experiments with diesel power. Buda four-cylinder diesel engines were installed in about 75 Model 80 row-crops and a few Model 80 standard-tread tractors. In 1940 a new smaller tractor, the Model 60, was introduced. The 60 was powered by a 121ci four-cylinder engine and had a 4-speed transmission. With 17 drawbar horsepower it was a marginal two-plow tractor, but it fit the needs of many small farms. The 60 shared the streamlined styling of the Model 70.

LEFT AND BELOW: The 90 was not restyled to match the rest of the Oliver line when the Fleetline series was introduced. Many, like the one below with a cable winch mounted on the back, were used in industry.

ABOVE: Early 88s used the same type of mesh grille as this 60 and the Model 70s.

After World War II a complete transformation of the tractor line was in order. The Model 80 had been a good tractor, but its design made it impossible to add electric starting. The entire tractor was scrapped in 1947 and replaced with the new, modern Model 88. The 88 used a six-cylinder Waukesha engine equipped for gasoline, distillate, or diesel fuel. The transmission had six forward speeds and two speeds in reverse. It was available in standard and row-crop chassis styles and was dressed in the streamlined sheet metal of the 60 and 70 tractors. Live hydraulics and PTO were available, firsts for an American-built tractor. Shortly after the 88 was introduced, it

BELOW: The Fleetline 88 was the star of the new line. With 34 horsepower, it was a strong three-four plow tractor.

BELOW: This simple clean grille design was used on all tractors in the Fleetline series.

BELOW: Dressed in stylish new Fleetline sheet metal, the 66 took the place of the old Model 60 in 1948. Live hydraulics and Oliver's Hydro-Lectric electrically controlled hydraulic system were major new features.

RIGHT: The OC-3 was one of the most popular small crawlers. It found work on farms, in industry, and on small construction jobs.

was restyled and reintroduced as part of the "Fleetline" series. This new series of tractors included the Models 77 and 66. In general design the 77 and 66 were similar to the Models 70 and 60 they replaced. The 77 offered all the features of the Model 88 in a 34hp, 3200-pound tractor. The Model 66 likewise offered these advanced features in a four-cylinder 25hp tractor weighing 2600 pounds. These three models not only shared common styling, but many of their mechanical parts were interchangeable. The use of many common parts allowed Oliver to offer tractors with superior features at prices that were competitive in the market. Oliver's big tractors, the Models 90 and 99, continued little changed.

In 1944 Oliver Corporation bought the Cleveland Tractor Company, a prominent manufacturer of crawler tractors. Oliver continued building

1948
Oliver 88 Row-Crop

After World War II the trend in row-crop tractors was toward more power and greater versatility. The Oliver 88 packed both into a streamlined package that brought real style to the farm tractor. Though it was manufactured briefly with sheet metal similar to the old Model 70, the Oliver 88 was first of Oliver's modern Fleetline series of tractors. The 88 offered many advanced features, including a modern six-cylinder engine, live power take-off, and live hydraulics. Engine options were a smooth six-cylinder diesel as well as a popular high compression gasoline engine, low compression distillate, and propane versions. Its fully enclosed engine, 6-speed transmission, and diesel power made the 88 a popular industrial tractor, but it was on the farm that the Oliver 88 made its mark. It was the first combatant in a row-crop tractor horsepower war that would last into the 1970s.

BELOW: The big powerful six-cylinder engine was built for Oliver by Waukesha. Electric starting and battery ignition were standard equipment on the Oliver 88.

LEFT: A 6-speed transmission, with three speeds in each of two ranges, also provided for two reverse speeds.

RIGHT: Oliver was a pioneer in providing a comfortable operator station. Steering was light, controls were within easy reach, and the padded seat made long days in the field less brutal.

RIGHT: The simple dashboard has water temperature, ampere, and oil pressure gauges. A light is provided for night operation. The light and ignition switch is on the right and the starter button is on the left.

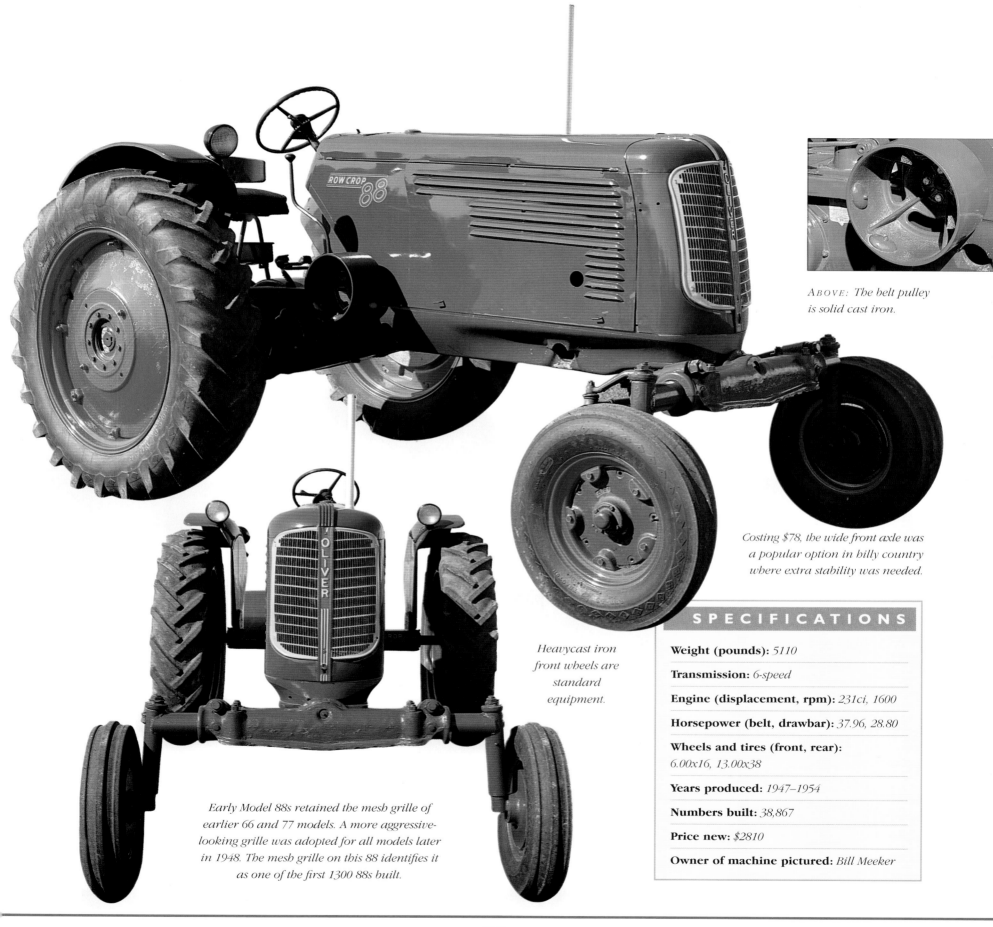

ABOVE: The belt pulley is solid cast iron.

Costing $78, the wide front axle was a popular option in hilly country where extra stability was needed.

Heavycast iron front wheels are standard equipment.

Early Model 88s retained the mesh grille of earlier 66 and 77 models. A more aggressive-looking grille was adopted for all models later in 1948. The mesh grille on this 88 identifies it as one of the first 1300 88s built.

SPECIFICATIONS

Weight (pounds): *5110*

Transmission: *6-speed*

Engine (displacement, rpm): *231ci, 1600*

Horsepower (belt, drawbar): *37.96, 28.80*

Wheels and tires (front, rear):
6.00x16, 13.00x38

Years produced: *1947–1954*

Numbers built: *38,867*

Price new: *$2810*

Owner of machine pictured: *Bill Meeker*

BELOW: The Super 55 was Oliver's first utility tractor— and it was a diesel! Gasoline power was also available, but the diesel option put Oliver a step ahead of the competition.

Cletrac design tractors under the Oliver-Cletrac name until 1965. Initially the Cletrac line was continued without even renaming the models. The 25hp Cletrac HG continued as a popular agricultural crawler as the Oliver-Cletrac HG.

The 1950s The tractor line continued with only minor changes. Liquefied petroleum (LP) gas and diesel became more popular as distillate lost favor as a tractor fuel. Oliver led the industry with small diesel tractors when it offered a diesel power option on the Model 66 in 1951. In 1952 the 99, little changed since 1937, received a facelift and a new engine. The tractor's 62 maximum horsepower came from a new six-cylinder diesel engine of 302 cubic inches displacement.

In 1954 Oliver offered its first utility tractor, the Super 55. Available with a 34hp diesel or gasoline engine and a draft-

ABOVE: A General Motors 3-71 two-cycle diesel gave the Super 99 GM 79 maximum horsepower.

BELOW: A fixed front axle and smaller rear wheels distinguished the 77 Standard from its more popular row-crop variant.

"New Oliver Super 55—Tops in versatility and quality at a cost surprisingly low."

– Oliver advertisement for the Super 55, "the tractor that has everything."

BELOW: A Super 66. Note the absence of engine side covers.

controlled three-point hitch, the Super 55 was a direct competitor to Ford and Ferguson utility tractors. The Model 66 was given more power and became the Super 66. The 77 and 88 got similar power increases and became the Super 77 and Super 88. Engine side covers, a hallmark of Oliver row-crop tractors since 1935, were dropped with the Super series. The 99 also received the Super styling, but was otherwise little changed. A new model, the Super 99 GM was added. This tractor was essentially a Super 99 powered by a General Motors supercharged, three-cylinder, two-cycle engine. Equipped with this engine, the tractor produced a maximum of nearly 79 horsepower.

The next new tractor from Oliver appeared at the other end of the power spectrum. With a scant 25 horsepower, the little one-row Super 44 was intended for vegetable, tobacco, and flower farmers who needed a small tractor with ample ground clearance and visibility. A few hundred were built in 1957 and 1958 before it was given a facelift and joined the new Oliver line-up as the 440.

After introducing an entire new line in the late 1940s and "Superizing" it in the mid-1950s, Oliver had no place to go with the latest update of its model line except to add numbers. The Hundred Series tractors, 440, 550, 660, 770, 880, and 990 were not exactly new. They received stunning new paint colors and sheet metal styling, and they were more powerful across the board, but for the most part they were dressed-up versions of old tractors. The first of the tractors to be upgraded in 1958, the 770 and 880 got a 2-speed Power-Booster transmission that allowed on-the-go shifting to a lower gear. With a more powerful

ABOVE: This Super 77 is outfitted for a day of working in the hot sun. These sun umbrellas were a nod toward operator comfort, helping make those long days in hot fields more bearable.

155ci engine the Super 55 became the 550 in 1958 also. The Super 66 got a partial facelift in 1959. Neither the 660 nor the 440 received newly styled sheet metal. The old sheet metal was merely painted the new colors. The 660 received the same engine as the 550 but the 440 made do with the same power and engine as the Super 44 it replaced.

There were big changes in the big tractors. The old Super 99, with the six-cylinder diesel engine, was restyled, retuned (to 67 horsepower), and renamed the 950. The Super 99 GM received the same treatment, now as the 990 putting out 84hp. A 995 Lugmatic was added to the line. This tractor was a 990 with five more engine horsepower and a torque converter ahead of the transmission. The torque converter enabled the tractor to pull through varying loads with greater ease, though the torque converter's inefficiency consumed the extra five horsepower. About 500 of the 990s were painted red, given Massey-Ferguson sheet metal and decals, and were offered by Massey-Ferguson as the Model 98.

LEFT: Under their new sheet metal the Hundred series of tractors, the 660, 770, and 880 were basically warmed-over Supers. The 770 and 880 got a 2-speed power shift for the transmission, but were otherwise not greatly changed.

BELOW: The 550 remained Oliver's only utility tractor and built on the success of the Super 55.

Like most manufacturers, Oliver spent the decade of the 1950s trying to remake their old designs quickly enough to keep up with the rapid changes in the market. Some changes they made were substantial, some were merely cosmetic, but all were destined to be obsolete as the decade of the 1960s dawned.

ABOVE: The Super 99 was available with a Waukesha six-cylinder diesel as well as the General Motors 3-71.

The 1960s On November 1, 1960, Oliver Corporation became a subsidiary of White Motors. With this acquisition, White turned one of America's best known and most innovative tractor manufacturers into little more than a farm machinery marketing company.

Good utility tractors were available cheap from overseas sources, so Oliver joined other tractor companies in importing tractors to fill that segment of the market. The Model 500, introduced in 1960, filled the market for a small (32 horsepower) utility tractor. The 600, first offered in 1962, was a 52hp utility model. The domestically built 550, at 40 horsepower, filled out the utility line.

White turned one of America's best known and most innovative tractor manufacturers into little more than a farm machinery marketing company.

ABOVE: Oliver imported its new, larger utility tractors from David Brown Industries of England.

A complete new line of tractors began to appear in 1960 with the 1800 and 1900 models. The 1800, Oliver's biggest row-crop tractor to date, delivered a maximum of 74 horsepower when operating on gasoline or LP fuel, and 70 horsepower with diesel power. It had a simple 6-speed transmission. The 1900 was powered by a General Motors 4-53 four-cylinder, two-cycle engine cranking out 92 horsepower. It was strictly a wheatland tractor. Series B versions of these tractors were unveiled in 1962. The Series B 1800 had a bigger engine and more power, plus a 2-speed "Hydra-Shift" power shift option and front-wheel-assist. The 1900 Series B had 99 horsepower as well as Hydra-Shift and optional front-wheel-assist. A replacement for the 880 was due by 1962. The 1600, with a 58hp six-cylinder engine offered all the features of the big 1800 in a four-plow tractor.

Oliver began offering three transmission options in the mid-1960s that would continue until the end of production. The standard transmission was either a 4- or 6-speed constant mesh gear drive unit. When equipped with Hydra-Shift, a 2-speed planetary transmission ahead of the standard

LEFT: The Oliver 1955, with 108 PTO horsepower, was available as two-wheel-drive. Front-wheel-assist was optional, though many farmers opted for the cheaper dual rear wheels.

three models. A few 1800s left in stock when the 1850 was unveiled were updated slightly and renamed the 1750.

The 770 was replaced in 1965 with the 1550. This tractor was basically a 1650 with a 53hp engine. Two years later the General Motors engine in the 1950 was replaced with a turbocharged version of the 310ci six-cylinder 1850 engine. In turbocharged form this engine produced 105 horsepower.

Oliver expanded its line of mid-sized tractors in 1965 by offering the Model 1250. This was a 35hp gasoline/39hp diesel utility tractor purchased from Fiat of Italy. In 1969 the gasoline option was dropped and the four-cylinder diesel was replaced by a three-cylinder diesel of similar capacity. The tractor became known as the 1250-A. The old domestically built 550 continued. A Model 1450, a 55hp diesel built by Fiat was also offered through 1968.

Bigger tractors were in order to expand the upper ends of the line and in 1968 the 2050 and 2150 were unveiled. Powered by a 478ci six-cylinder diesel designed by White, the 2050 produced 119 horsepower. The 2150 has a turbocharged version of this engine and produced 131 horsepower. These tractors were available with a 6-speed transmission and Hydra-Power to give 18 forward speeds. Front-wheel-assist, as well as an adjustable front axle for row-crop work, were optional.

The 1970s The tractors imported from Italy were Olivers only by virtue of being painted Oliver colors. White began adding its name on imported models when the Italian line was upgraded in late 1969. The 1250-A became the White-Oliver 1255, and the 1450 was replace by the 1355, essentially a 1255 with another cylinder added to its three-cylinder diesel engine. The

transmission, the operator could shift between low and direct ranges in any gear without stopping. The Hydra-Power option was a low-range/direct-drive/high-range transmission mounted ahead of the standard transmission that could be shifted under power. It allowed shifting from an underdrive, to direct drive through the Hydra-Power unit, to overdrive, without stopping the tractor, giving the 6-speed standard transmission 18 speeds.

In February 1962 White Motors bought Cockshutt Farm Machinery Company of Brantford, Ontario, Canada. Production of Cockshutt tractors was terminated. Cockshutt dealers in the United States were forced to sell Oliver tractors, and Oliver dealers in Canada were forced to sell red-painted Oliver tractors with the Cockshutt name emblazoned on the hood.

By 1964 the upper end of the tractor line was falling behind. A new series, the 1650, 1850, and 1950, was introduced with more horsepower and a few other changes. The gasoline 1650 now put out 67 horsepower, the gasoline 1850 produced 92, and the 1950, with its GM Detroit diesel engine screaming at 2400rpm, produced 106 horsepower. The 1950 was available as a row-crop tractor in addition to the standard-tread and front-wheel-assist models. Hydrostatic front-wheel-assist was available for all

*The name Oliver will be remembered as one of
the greats in American agricultural history.*

BELOW: The 1655 was the smallest domestically built Oliver.

next year the 1255 was given a larger engine. White sold it under several names, including White, Oliver, Cockshutt, and Minneapolis-Moline.

Larger tractors were still domestically produced. The 1550 was upgraded to the 1555 with the addition of a new grille. The 1650 became the 1655 with the addition of the new grille and a retuning of the engine to give about six per cent more power. Front-wheel-assist was available as an option. The 6-speed transmission was available with Hydra-Power. The 1750 likewise got an eight per cent power boost and a new grille to become the 1755. Both the gas and diesel version produced 87 horsepower. The 1855 was a retuned 1850, as was the 1955.

Oliver tractors received the ultimate insult in 1971. White Motors began painting tractors designed and built by Oliver's one-time rival Minneapolis-Moline in Oliver colors and putting the Oliver name on them. The Oliver 1865 was a repainted Minneapolis-Moline G-950. It was a 98hp tractor that offered features similar to those of the 1855. The 2055 was essentially an 1865 with a 110hp engine. White's crossbreeding scheme began to make sense in the larger sizes, as Oliver had nothing to compete with the 2155, a 141hp Minneapolis-Moline G1350. The 2455

added a 151hp articulated tractor to the Oliver line for the first time. The largest of the line was the articulated 169hp Model 2655.

White made a reversal of policy in 1972 and began to build Minneapolis-Moline tractors in Oliver's old Charles City, Iowa plant using Oliver components. The last Oliver tractor came out of the plant in 1976. It was a Model 2255, a 147hp tractor with a Caterpillar engine.

The End of the Line   The rich history of the Oliver Farm Equipment Company and the memories of Joseph Oliver, Charles Hart, and Charles Parr may have been better served if White Motors had simply stopped using the Oliver name when it bought the company in 1960. But the decades of being an innovator and builder of quality farm machinery far outweigh the ignominy of the later years, and the name Oliver will be remembered as one of the greats in American agricultural history.

BELOW: When equipped to burn diesel fuel, the 1855 shared the same basic engine as the 1955, tuned to produce 97 horsepower. A 92-horsepower gasoline engine was also available.

CHAPTER SIX

THE AMERICAN TRACTOR

★

INTERNATIONAL HARVESTER

ABOVE: The red and black IH logo was created by the famous designer Raymond Loewy in the early 1940s.

The International Harvester Company (IHC) was formed in 1902 by the consolidation of McCormick Harvesting Machine, the Deering Harvester Company, and three smaller farm equipment manufacturers into one huge corporation. IHC built its first tractors in 1906, just four years after Hart-Parr built its first successful tractor. They were assembled in Akron, Ohio under contract by Ohio Manufacturing. These early tractors consisted of IHC Famous stationary engines mounted on Morton frames with Morton steering, wheels, and gearing. They were produced in 10-, 12-, and 15-horsepower sizes. All had a single speed forward with friction disk drive. This first tractor venture was such a success that the company scheduled 200 more for production in 1907.

Friction drive was adequate for low horsepower tractors used on belt work, but it quickly became obvious that a sturdier drive mechanism was need for drawbar work. Improved tractors with gear drive were designed and construction began in 1908. The success of its early tractors gave IHC the confidence to bring tractor production in-house, to IHC's Milwaukee Works. In 1910 gear drive tractors of 20, 25, and 45 horsepower were produced there under the Reliance brand name. In less than a year the name was changed to Titan.

Divided Loyalties Though now part of the same company, bitter rivals McCormick and Deering maintained separate dealers, distributors, and product lines. Often there would be a McCormick and a Deering dealer in the same town, with Titan tractors being sold by Deering dealers.

McCormick dealers demanded that IHC provide them a tractor to sell and they didn't want a repainted, renamed Titan. The Mogul line of tractors was the result. Though there are exceptions, for the most part Titans were designed and built in Milwaukee, used chain steering, and hit-and-miss governed Famous engines. Mogul tractors were designed and built at IHC Tractor Works in Chicago, used automotive steering, and were powered by Mogul throttle-governed engines.

From a business standpoint this dual line my have been inefficient, but it greatly benefited the tractor industry as a whole. The development of the farm tractor was in its virtual infancy at this time. The fierce competition between these two well-financed rivals to build the better machine accelerated the development of dependable tractors and brought the day closer when the tractor would replace the horse.

Through 1917 IHC produced small quantities of a large variety of tractors. Production runs were typically a few hundred units. Power ranged from 25 to 60 belt horsepower while weight ranged from 10,000 to over 22,000

ABOVE: International took a great leap forward in tractor design with the Type A Gear Drive tractors.

"This new Mogul 8-16 tractor will do the work of eight horses in the orchard. Being a four-wheeled, all-purpose tractor, you can use it every working day." – Company advertisement, 1915

pounds. In the mid-1910s, quietly and with little fanfare, IHC unveiled two small tractors. These, the 4900-pound 8-16 Mogul and the 5700-pound 10-20 Titan, would lead the way to a new era in tractor farming. The little Mogul went into production in 1914 and in one year 8-16 Mogul production surpassed 5000 tractors, more than all IHC tractors of all other models combined. The 10-20 Titan followed a year later with similar

RIGHT: The success of the 10-20 Titan convinced International to get out of the big gas traction engine business.

BELOW: The 2-speed Model 8-16 was International's first successful small tractor.

success. The writing was on the wall, and IHC suspended production of big tractors in 1917. The next year 17,675 10-20 Titans were sold.

The era of the lightweight tractor had arrived. The 8-16 Mogul, with its single-cylinder, hopper-cooled engine and a 1-speed planetary transmission

was improved with a closed cooling system and a 2-speed transmission. The 10-20 Titan, whose design was completed only one year later, took advantage of lessons learned with the 8-16 and featured a two-cylinder engine, 2-speed spur gear transmission, and a closed cooling system from the beginning.

In 1918, as a result of anti-trust action by the United States Justice Department, IHC consolidated its McCormick and Deering dealerships. Henceforth each sales territory would have only one IHC dealer and all IHC tractors were to be called Internationals.

A more modern tractor, the International 8-16, was inaugurated in 1917. Though it was initially intended to be part of the Mogul line, the 8-16 was the first International tractor. It had three transmission speeds, a modern four-cylinder engine that turned an impressive 1000rpm, and weighed 3660 pounds. Unlike the little Titan and Mogul, the new tractor had modern radiator cooling. The radiator was mounted in a unique location, behind the engine over the flywheel. An innovative feature was its power take-off, a first for an American production tractor. It was a successful tractor, but retained such weaknesses as a riveted frame and open drive to the rear wheels.

1924
McCormick-Deering 15-30

Nothing exemplifies the rapid advance of tractor technology more than a comparison of International's 1919 Titan 15-30 and the same corporation's McCormick-Deering 15-30 unveiled two years later. The 1919 tractor was over 13 feet

long, weighed 8700 pounds, and cost $2300. The new 15-30 could do the same work, yet it was only 11 feet long, weighed 5750 pounds, and cost $1250. What the data do not show is the incredible simplification of the tractor. The 1919 tractor had seven carburetors. The engine was exposed in front of the operator so he could perform the frequent adjustment and lubrication chores it required. Once it was running, the engine of new 15-30 needed little attention. It was under a sheet metal hood far from the operator. The McCormick-Deering 15-30 was a leap forward in tractor technology that put four-plow power in the hands of farmers who previously found tractors too complex and expensive to consider.

LEFT: Replaceable steel lugs gave the 15-30 plenty of traction in most field conditions. Few 15-30s are seen with wheel extensions.

RIGHT: Compact and rugged, the 15-30 was a favorite among wheatland farmers, as well as custom threshermen, throughout the 1920s and early 1930s.

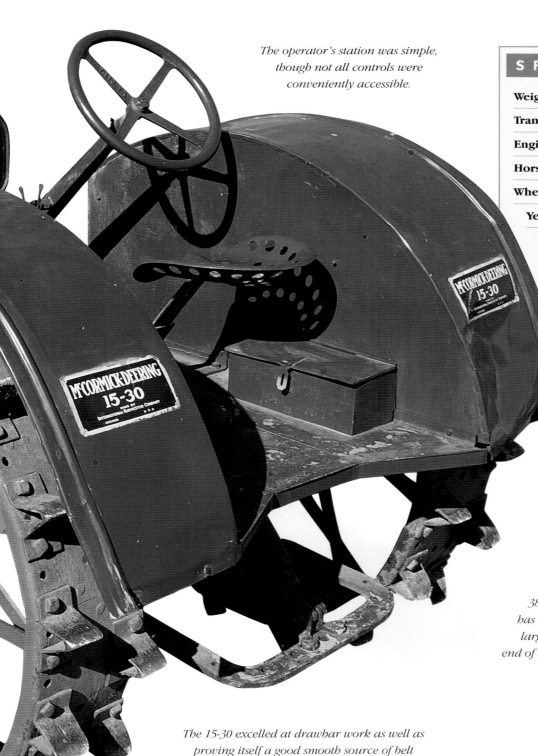

The operator's station was simple, though not all controls were conveniently accessible.

SPECIFICATIONS

Weight (pounds): 5750

Transmission: 3-speed

Engine (displacement, rpm): 381.7ci, 1000

Horsepower (belt, drawbar): 30, 15

Wheels and tires (front, rear): 34x6, 50x12

Years produced: 1921–1929

Numbers built: 128,125

Price new: $1250

Owner of machine pictured:
Henry Shriver

BELOW: Kerosene fuel tended to get past the piston rings and dilute the motor oil. Frequent oil changes were needed.

ABOVE: Though the 382ci four-cylinder engine has only two main bearings, large ball bearings at either end of the crankshaft, it proved to be a reliable design.

RIGHT: This owner cut a hole in the side curtain for easy access to the carburetor.

The 15-30 excelled at drawbar work as well as proving itself a good smooth source of belt power. It was simple to operate and required relatively little maintenance.

1927
Farmall

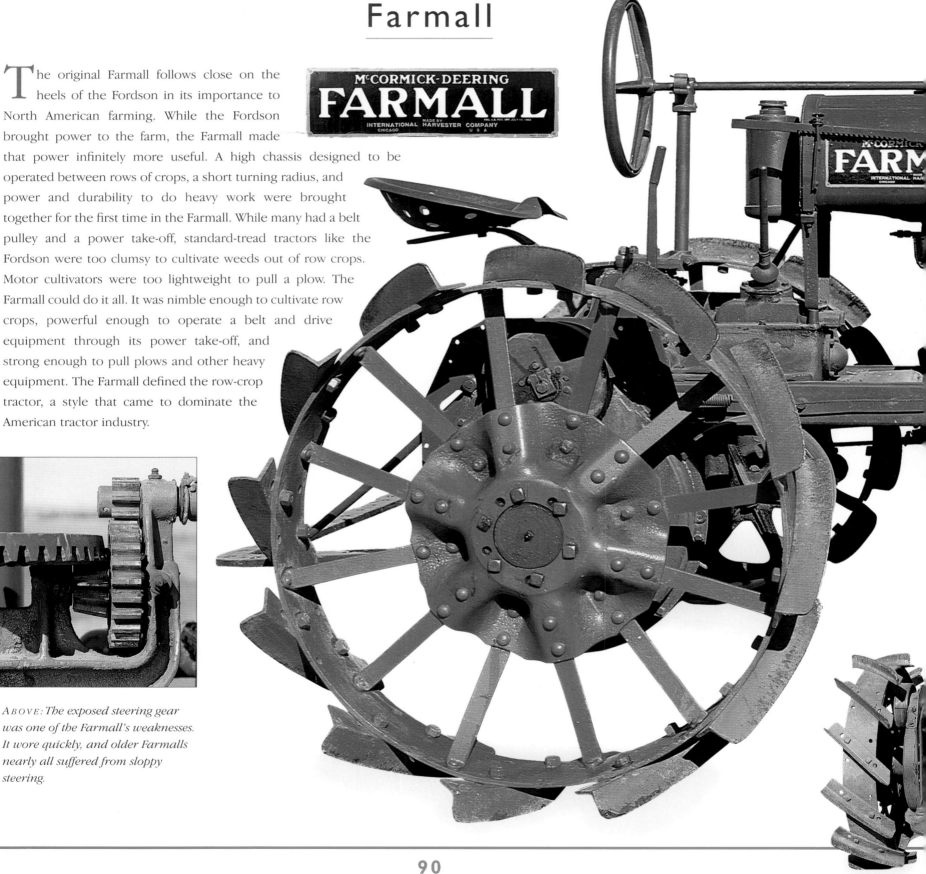

The original Farmall follows close on the heels of the Fordson in its importance to North American farming. While the Fordson brought power to the farm, the Farmall made that power infinitely more useful. A high chassis designed to be operated between rows of crops, a short turning radius, and power and durability to do heavy work were brought together for the first time in the Farmall. While many had a belt pulley and a power take-off, standard-tread tractors like the Fordson were too clumsy to cultivate weeds out of row crops. Motor cultivators were too lightweight to pull a plow. The Farmall could do it all. It was nimble enough to cultivate row crops, powerful enough to operate a belt and drive equipment through its power take-off, and strong enough to pull plows and other heavy equipment. The Farmall defined the row-crop tractor, a style that came to dominate the American tractor industry.

ABOVE: The exposed steering gear was one of the Farmall's weaknesses. It wore quickly, and older Farmalls nearly all suffered from sloppy steering.

SPECIFICATIONS

Weight (pounds): *3650*

Transmission: *3-speed*

Engine (displacement, rpm): *220.9ci, 1200*

Horsepower (belt, drawbar): *20.05, 13.27*

Wheels and tires (front, rear): *25x4, 40x6*

Years produced: *1924–1932*

Numbers built: *135,000*

Price new: *$950*

Owner of machine pictured:
Henry Shriver

LEFT: With the drawbar and belt pulley removed, the Farmall offered 30 inches of clearance over two rows of crops.

The Farmall looks so simple—little more than an engine and transmission in a crude frame—but the configuration of that frame made the tractor revolutionary. Those cables running the length of the frame rail are an automatic system of applying the turning brakes when the steering wheel is turned.

ABOVE: At about the time the Farmall came along, engineers were realizing that clean combustion air was essential for long engine life. The Farmall's air intake was mounted high up and far forward to keep it out of the dust raised by the tractor.

RIGHT: Though it was a totally new concept in tractors, the Farmall, introduced in 1924, remained in production with only minor changes for 14 years. The F-20 of 1932 was the only significant update.

Tractor development proceeded at a feverish pace at IHC. The success of the Wallis Cub with its one-piece frame was not lost on company engineers, nor was the farmer's cry for a general purpose tractor. In the late teens IHC began work on a tractor so exceptional that it would carry the company for a decade, and a decade was an eternity in the fast-paced tractor industry.

The 1920s The first new machine of the decade was introduced into the teeth of the depressed postwar economy of 1921. It was the remarkable new International 15-30 gear-drive tractor. It was powered by a compact, four-cylinder vertical engine with pressure oiling and overhead

valves. IHC had such high regard for its engine that it offered a lifetime guarantee on the crankshaft. The engine was mated to a sealed, 3-speed spur gear transmission that drove a sealed, gear-drive rear axle. The power train was mounted in a one-piece cast iron frame that kept all components in perfect alignment. The new tractor weighed only 5750 pounds and easily pulled three plows. It was simple enough that it did not require an engineer to operate it, and at $1250 the average size farmer could afford it.

The always-squabbling McCormick and Deering interests were unhappy with the new tractor name, so in 1923 McCormick-Deering replaced the name International on IHC tractors. In more important news, a two-plow version of the 15-30, the 10-20, was introduced. The McCormick-Deering 10-20 incorporated all the advanced features of the 15-30 in a 3700-pound tractor. Though not known by the public, another even more important event took place that year. The McCormick-Deering Farmall was approved for production.

Prior to the introduction of the Farmall, a tractor's usefulness was limited to what it could pull with its drawbar or drive by the belt. About 1910, engineers began to think about using tractors to cultivate row crops. Tractors specially designed for cultivating were effective at that task, but too light and underpowered to be much good at draft or belt work. What was needed, and what IHC engineers began working on in 1916, was a machine with the power of a heavy tractor and the agility of a horse. After seven years of experimentation the engineers at IHC had it: the Farmall. The aptly named Farmall—it truly could do it all—was a revolutionary machine. With 13 drawbar horsepower and 3650 pounds, it was powerful and heavy enough to pull a two-bottom plow. Its 20 belt horsepower was adequate for driving threshers and shredders, and its power take-off shaft allowed it to drive power binders and mowers. With narrow front wheels and turning brakes it was nimble enough to turn in its own length. Its high clearance and wide wheel spacing allowed it to drive between rows of corn or cotton to cultivate weeds with specially designed mounted cultivators.

The Farmall was a hit with farmers, but was viewed with skepticism within the IHC organization. The 10-20 was selling well and making a profit. The new Farmall had similar power and sold for a similar price, but

ABOVE AND RIGHT: The 10-20 and Farmall F-20 were International's most popular models. The cutaway view of the 10-20 at the right illustrates graphically just how simple the tractor was.

"If it isn't a McCormick-Deering, it isn't a Farmall"

– Company advertisment, 1934

The 1930s The Great Depression was deepening as the new decade opened. Like all business, the agricultural tractor industry suffered. But fewer tractor companies failed in the first half of the 1930s than in the early years of the 1920s. Three factors contributed to the relative strength of the industry: Many marginal companies had been weeded out in the industry decline of the early 1920s, Henry Ford's exit of the business in 1928 opened up the market, and for the farmer, the investment in a tractor was just too profitable to resist.

IHC entered the decade with three tractors in the line. The big 15-30 had just been updated and the 10-20 was selling so well it needed no changes. Though the Farmall was still in its first generation, 75,000 were in the field and much had been learned about the needs of row-crop farming. Every major tractor manufacturer had followed the Farmall closely and noted its weaknesses. John Deere, Allis-Chalmers, Case, Oliver, and Minneapolis-Moline all had more powerful row-crop tractors in the field or on the drawing board.

Perhaps IHC was watching the competition too closely and not listening to farmers closely enough. While farmers needed a smaller Farmall, IHC's first new tractor in seven years was a larger version of the row-crop tractor. The Farmall F-30 incorporated all the Farmall's attributes in a 5300-pound 30hp tractor. While it was a good machine, for farmers the F-30 was the wrong tractor at the wrong time.

New Models In 1932 the second-generation Farmall, the F-20, replaced the original Farmall. It had slightly more horsepower and much improved steering, but retained the look and size of the Regular. While it addressed some of the weaknesses of the Regular, the F-20 did nothing to open up new markets. The thousands of farmers who only had one hundred acres to till still needed a smaller row-crop tractor. In 1932 IHC gave them just what they wanted, the F-12.

This small row-crop tractor weighed only 2700 pounds. It had 10 drawbar horsepower, enough for a one-bottom plow, and 15 belt horsepower. Like the other Farmalls, it could fit between rows and was an

ABOVE: The Farmall F-12 was introduced in 1932 to fill the needs of thousands of small American farms. The wide front axle on this tractor was uncommon. Most F-12s were sold with a single steel front wheel.

ABOVE RIGHT: International had to convince owners of small farms that a tractor could replace a team of horses, and make them more money. The F-12 was versatile and inexpensive enough to do both.

production was so low it was losing money. Corporate money men were afraid the money-losing tractor would take sales away from the money maker, so they limited production and allowed the Farmall to be marketed only in areas where the 10-20 wasn't selling well. Cooler heads prevailed, however. Though few in management realized it, from the time the Farmall was unveiled IHC's only real problem was how to produce them fast enough. Sales rose from 200 tractors in 1924 to 4430 in 1926, the year an entire manufacturing plant was dedicated to the tractor. Farmall sales in 1930 exceeded 40,000 tractors.

The 10-20, the 15-30, and the Farmall comprised the entire IHC tractor line after 1924. The two smaller tractors remained largely unchanged, but the 15-30 was improved significantly just before the decade ended. Changes in the engine increased power output to 22 drawbar horsepower and 36 horsepower at the belt. The tractor was popularly called the 22-36, though to IHC it was the Model 15-30 that put out 22-36 horsepower.

"The First Day with a Farmall
A Red-Letter Day on any Farm"
– International Harvester slogan, 1930s

BELOW: The W-12 was a standard-tread tractor built around an F-12 engine.

LEFT: Big crawler tractors were required to farm the vast acreages and steep hills of the wheatlands of Washington State. With 34 horsepower, the T-40 TracTracTor could handle the biggest tillage and harvest equipment.

RIGHT: The WD-40 was America's first diesel-powered wheel tractor. It produced 49 maximum belt horsepower.

IHC also built a line of crawler tractors that were used in agriculture. The crawlers were based around the 20-, 30- and 40hp wheel tractor engines and gave similar horsepower figures.

The F-12 was updated in 1938 with a better steering wheel position and a retune of the engine. The upgraded tractor was called the F-14 and produced 12 drawbar and 15 belt horsepower.

excellent tractor for cultivating two rows at a time. A new feature was rear wheels whose width could be adjusted by sliding them on the axle.

The venerable 15-30 was replaced in 1932 with a standard-tread tractor built around the F-30 engine. At 20-31 horsepower, the W-30 was slightly less powerful, but was cheaper to produce. The W-12, a little standard tractor built around the F-12 engine, joined the line in 1934. The 10-20 continued unchanged until 1939.

A new line of big standard-tread tractors was developed for the western wheatlands. The first of these, the WD-40, was also the industry's first mass-produced diesel tractor. It weighed 7550 pounds and its four-cylinder diesel engine produced 28 drawbar and 44 belt horsepower. The W-40, a gasoline version, used a six-cylinder IHC truck engine and produce the same power figures. This six-cylinder tractor equipped to burn distillate was known as the WK-40 and produced slightly less horsepower.

Design By Loewy IHC had begun work on a totally new line of tractors in 1936. For the new tractors IHC threw out everything but the name and started clean. The company secured the services of the Raymond Loewy design studio to apply the concepts of Industrial Design to its new farm tractors. The Industrial Design movement had been successful at taking everyday utilitarian machines and making them more attractive to look at while retaining their usefulness. In designing the tractors, Loewy enclosed the fuel tank, steering bolster, and radiator in a single streamlined housing. He hired an orthopedic

RIGHT: W-12s were most often seen as industrial tractors or with orchard fenders. A golf course version was also available. Though it was available in America, F-12s with the wide front axle option were sold mostly in Europe.

INTERNATIONAL

Model W-12
Tractor Power to Fit the Smaller Diversified Farms

WHAT THE W-12 WILL DO

INTERNATIONAL

FARMALL 12
The Original All-Purpose Tractor for the Smaller Farms

WHAT THE F-12 WILL DO

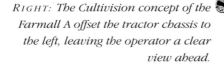

surgeon to shape the seat, and designed the wheels to combine the impression of strength with lightness.

The first of the new tractors, the Model A, was introduced in July 1939. This one-plow tractor was the replacement for the F-14, Unlike the F-14 it had an adjustable wide front axle and the rear wheels, though adjustable, were not mounted on sliding axles. The engine was a 113ci four-cylinder overhead-valve unit. Rated horsepower was 13 drawbar and 16 belt. Called the Model A "Cultivision," it was especially designed to cultivate a single row. A Model B was offered for cultivating two rows. It was actually just a variant of the Model A with a tricycle-type front end and longer left rear axle.

Farmers waited until August to see the F-20 and F-30 replacements. These, everyone knew, would be the big sellers. The Farmall H, with 19 drawbar and 24 belt horsepower, was

BELOW: Steel wheels were always available for the Farmall M, though the tractor was designed for rubber tires. Few were ordered on steel until the rubber shortages of World War II forced buyers to take steel or nothing.

RIGHT: The Cultivision concept of the Farmall A offset the tractor chassis to the left, leaving the operator a clear view ahead.

RIGHT: High-crop versions of both the H and the M, called the HV and MV, were available after 1942. This is an HV.

LEFT: The Model F-20 by Ertl is a perennial favorite at Christmas time.

BELOW: A few of the prototype Farmall Ms had white-painted grille bars. Photos of these made it into advertising brochures, though no production Ms ever had the distinctive feature. The H and M were the most popular International tractors ever, and thousands are still at work on American farms.

the successor to the F-20. It was available with a distillate engine that produced 17/21 horsepower. The H incorporated narrow front wheels, adjustable rear wheel track, steering brakes, and operator comfort into an outstanding tractor. Farmers used to the cramped quarters on an F-20 must have felt they had moved to the big house on the roomy platform of the H.

The 33-belt horsepower M was the three-plow tractor, and initially suffered from the legacy of the F-30. But farmers soon learned what Raymond Loewy had known. With proper design, even a big heavy tractor can be a breeze to

RIGHT: Farmall had a hit on its hands with the little Cub. This small 8hp tractor found work everywhere from farms to suburban estates.

operate. They discovered that the M was more nimble and easier to operate than the little old F-12 and could do four times as much work. The M was available with a distillate, gasoline, or a diesel engine.

New standard-tread tractors were also added to the line. At the top, the completely new 6000-pound W-9 was powered by a 335ci four-cylinder engine tuned for

ABOVE: The Farmall B shared the engine and transmission of the A. It differed in having narrow front wheels and a centrally mounted chassis and controls.

1947
Farmall H

The Farmall H was perhaps no more capable than other tractors of its size, but it did its job with unmatched ease and economy. When it came on the market in 1939 the H perfectly fit the power needs of the tens of thousands of 250-acre farms in America. International Harvester's most popular tractor throughout the 1930s had been the F-20, but advances by other manufacturers had made the heavy, incredibly uncomfortable F-20 obsolete. The Farmall H produced similar power, but was lighter and infinitely more comfortable and easy to operate. It was modern and streamlined, and most important, it came along when over 200,000 Farmall Regulars and F-20s were nearing the ends of their useful lives. And at a price under $1000, the H was affordable. Over 300,000 Hs were sold between 1939 and 1953 and, due to its utility and durability, many of them can be still found at work on America's farms.

LEFT: The new two-color IH logo was designed by Raymond Loewy and introduced in 1944.

SPECIFICATIONS

Weight (pounds): *5375*

Transmission: *5-speed*

Engine (displacement, rpm): *152.1ci, 1650*

Horsepower (belt, drawbar): *20.69, 16.99*

Wheels and tires (front, rear):
5.50x16, 10.00x36

Years produced: *1939–1953*

Numbers built: *390,317*

Price new: *$962*

Owner of machine pictured: *Lauren Secor*

BELOW AND BELOW LEFT: Harvester didn't leave important information to chance. Gear locations were cast into the transmission, and oil filter maintenance instructions were set right on the filter.

RIGHT: It took a good pair of eyes to see the oil pressure and water temperature gauges. They were located in the top of the hood about four feet from the operator.

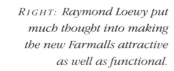

RIGHT: Raymond Loewy put much thought into making the new Farmalls attractive as well as functional.

RIGHT: The operator's station was comfortable by the standards of the day, and most controls were easily accessible.

The new IH logo and absence of "Deering" on the hood decal identify this tractor as a postwar model. Two fuel tank caps indicate this tractor was equipped to burn distillate fuel.

★

Powershifting was such an advance that
within 20 years nearly every tractor had it.

LEFT: The W-6 used the same engines as the Farmall M. Like the M, the W-6 could be equipped for distillate, gasoline, or diesel fuel. The W-6 was also the basis for the O-6 and OS-6 orchard tractors.

1953. The M sold well too, shipping well over a quarter of a million tractors. IHC built several variations of the A, H, M, and W series to adapt them for high clearance, grove, and rice field work, as well as building crawler tractors based on their mechanical components.

For some farmers, even 16 horsepower was too much, so IHC gave them the Cub in 1947. With 7 drawbar horsepower and only 8 belt horsepower (not that it was used much for belt power)

either gasoline or distillate fuel. On gasoline it produced 35/45 horsepower and on distillate it produced 33/42 horsepower. A diesel engine in the same chassis was offered as the WD-9. This engine had the same gasoline starting system as the M diesel. Standard tractors and orchard variants based on the M and H engines were also offered.

the tiny Cub would seem too small to be of much use. But, it found dozens of uses on the farms in the southeast and immediately began to outsell the Model A.

In 1948 the Model A was improved by replacing the pneumatic lift system with a hydraulic system. This improved A was called the Super A. The Model B, intended to be a two-row cultivator tractor, had a serious limitation in that application. The range of adjustment for the rear wheel width was insufficient, making it unsuited to many applications. IHC addressed the issue in 1948 with the Farmall C. The C used the same engine and transmission as the Farmall A with a centrally mounted chassis and a

A Best-Seller All of the new IHC tractors were immensely popular, but the H outshone them all. IHC built over 391,000 Model H tractors before the first major update in

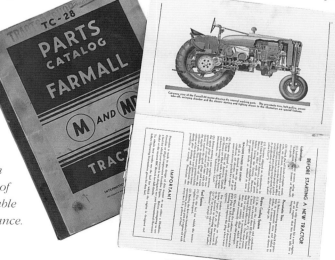

RIGHT: The cutaway illustration illustrates the essential simplicity of the Farmall M. Most farmers were able to perform their own maintenance.

tricycle front end with large-diameter rear wheels sliding on the rear axles. The B was dropped from the line. Farmers found the C more to their liking and put it to work on light chores. About this time the "Deering" was dropped from the brand name, and the tractors became McCormick Farmalls.

RIGHT: International tried to interest the industry in its own two-point hitch. This proprietary system, called Fast-Hitch, was a good idea that was overwhelmed by the universal popularity of the three-point hitch.

LEFT: The Farmall C combined the engine of the A with a tricycle front end and large diameter rear wheels to provide a small row-crop tractor.

The tractor line changed little until 1951, when the Farmall Super C debuted. IHC gave the Model C a larger bore, boosting horsepower to 16/21. In 1952 the H and M followed suit, receiving a horsepower boost and the names Super H and Super M. The W series tractors based on the M and H also received a promotion to Super W-4 and Super W-6.

1954
Farmall Super MTA

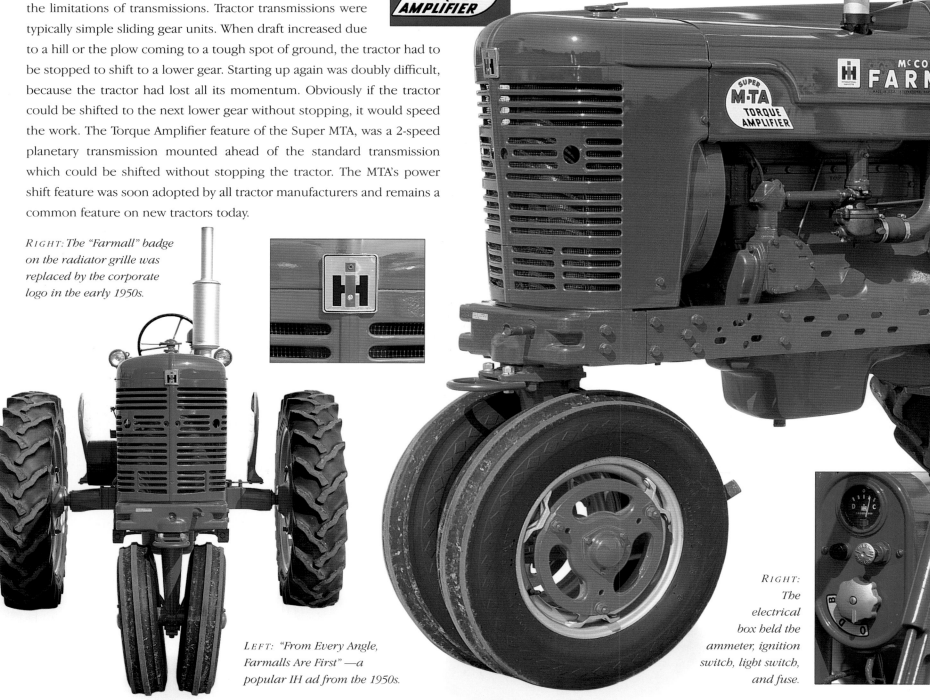

With durable engines, diesel power, live power take-off, and hydraulic systems becoming common by 1954, the next step in tractor development was to address the limitations of transmissions. Tractor transmissions were typically simple sliding gear units. When draft increased due to a hill or the plow coming to a tough spot of ground, the tractor had to be stopped to shift to a lower gear. Starting up again was doubly difficult, because the tractor had lost all its momentum. Obviously if the tractor could be shifted to the next lower gear without stopping, it would speed the work. The Torque Amplifier feature of the Super MTA, was a 2-speed planetary transmission mounted ahead of the standard transmission which could be shifted without stopping the tractor. The MTA's power shift feature was soon adopted by all tractor manufacturers and remains a common feature on new tractors today.

LEFT: Farmall's torque amplifier was first in an industry-wide move toward shift-on-the-go transmissions.

RIGHT: The "Farmall" badge on the radiator grille was replaced by the corporate logo in the early 1950s.

LEFT: "From Every Angle, Farmalls Are First" —a popular IH ad from the 1950s.

RIGHT: The electrical box held the ammeter, ignition switch, light switch, and fuse.

The exposed wires, hoses, and levers required by modern lights, torque amplifier, and hydraulic systems introduced clutter to Loewy's original open, clean design.

SPECIFICATIONS

Weight (pounds): 5725

Transmission: 10/2-speed

Engine (displacement, rpm): 264ci, 1450

Horsepower (belt, drawbar): 41.28, 33.18

Wheels and tires (front, rear):
6.00x16, 13.00x38

Years produced: 1952–1954

Numbers built: 22,000

Price new: $2925

Owner of machine pictured:
Marvin Roblena

BELOW: The Super MTA was clearly the end of the line for the M design. Pipes and hoses were sprouting up everywhere, and needed to be integrated more closely into the tractor.

LEFT: Though it looked much like the Farmall M of 1939, the Super MTA was a much more sophisticated tractor.

RIGHT: The Farmall 450 was the ultimate incarnation of the Farmall M. It differed primarily in having increased operator comfort and greatly improved hydraulics.

BELOW: Unlike the gasoline-start American Farmall M, the BMD had a direct start diesel engine.

In mid-1954 IHC began unveiling a line of improved tractors based on the models introduced in 1939. The Super A received a slight bore increase and increased power and became the 100. The Super C used the same engine in a slightly higher state of tune and became the 200. An ultra-low transmission speed called the Hydra-Creeper was available. The Farmall H received new sheet metal and was significantly improved with a Fast Hitch, live PTO, and Torque Amplifier and became the 300. The engine was enlarged to 169ci and the tractor was rated at 34 PTO and 27 drawbar horsepower. The 300 was available with equipment to burn gasoline, distillate, or LP fuel. The M got similar new looks and a slight power increase and became the 400. It too was available with Fast Hitch, live PTO, and Torque Amplifier and could be equipped to burn gasoline, distillate, LP, or diesel fuel. W-300 and W-400 standard-tread tractors based on the 300 and 400 were added to the line. A 600 based on the WD-9 was also in the line-up.

Technical Improvements

Technical Improvements IHC tractors were falling behind technologically. The M&W Gear Company had developed a 9-speed transmission as well as a live power take-off for the H and M that had proven popular and successful. Farmers and aftermarket suppliers were forging ahead with better Farmalls even if IHC wasn't.

In 1953 the Super MTA was unveiled to address this. The "TA" stood for "Torque Amplifier," a 2-speed planetary transmission that fit ahead of the regular 5-speed. The Torque Amplifier allowed farmers to shift between two ranges without stopping the tractor. Powershifting was such an advance that within 20 years nearly every tractor had it. The TA model also came with a live power take-off. The TA was available on the Super M and Super W-6 models. Also in 1953 IHC introduced the Fast Hitch on the Super C. The Fast Hitch was IHC's answer to Ferguson's three-point hitch. It allowed specially designed plows and other rear-mounted implements to be attached with a minimum of effort.

A brand new model was introduced in 1955, the International 300U. This little tractor combined the 300 engine and many of its features in a low-platform tractor that competed with the market virtually owned by Ford and Ferguson.

A Change of Image The tractor line received two more facelifts before the end of the decade. In 1956 the grilles were painted white, a large white field was added to the hood sides, and the tractors all got new model names. The 100 became the 130, the 200 went to the 230, the 300 and 300U became 350 and 350U, the 400 became the 450, and the 600 became the 650.

ABOVE: The Farmall 340 was the last of a dying breed. Soon small row-crop tractors would be supplanted by utility tractors.

In late 1958 the full-line was changed yet again, but this time there was big news. New squared-off sheet metal was designed for the full line and even more white paint was used on the hoods. The smaller tractors were renamed

> *"Get square dance handling ease...plus sod-bustin' plow-power! with a McCormick Farmall 230."*
> — *Magazine advertisement for the Farmall 230, 1960*

again. The Cub was still the Cub, but the 130 became the 140. The 230 became the 240. A new small row-crop, the 340, with a 135ci engine joined the line bringing the Torque Amplifier to a small tractor for the first time.

The two big Farmalls were replaced by the 460 and the 560. These tractors were powered by gas, diesel, or LP versions of the six-cylinder engine used in the TD-9 crawler. The diesel 460 developed a maximum of 50 PTO horsepower while the 560 put out 59. Integrated hydraulic systems, three-point hitches, and added attention to operator comfort brought them up to date with the most modern tractors on the market. The final drive, however, was decidedly not up to date. It was, in fact, the same unit originally designed for the 31hp Model M in 1938. The drives soon began to fail in the field. International spent over $100,000 revising and replacing final drives and by the time the tractors had been made right, International's reputation was tarnished irreparably.

Utility versions of the 240, 340, and 460 were offered, as well as a big wheatland tractor based on the 560, the 660. The 660 fell victim to final drive failure too, and added to International's woes. Product failure was not the only problem IHC faced. John Deere's sales surpassed International's for the first time in history.

ABOVE AND LEFT: The International 560 put the engine and transmission of the Farmall 560 in a standard-tread tractor. Dual Hydra-Touch hydraulics and live PTO were optional equipment.

The 1960s Shaken but undeterred, International plunged into the new decade with designs for an all-new line of tractors. First on the agenda was a four-wheel-drive tractor to compete with the likes of Steiger and Wagner, who were building high-horsepower wheel tractors to replace big crawlers in the west. The result was the International 4300. They never sold in large numbers, nor were they big money makers, but with 300 engine horsepower these large tractors kept the International name near the top of the horsepower list.

1961
Farmall 560

In the postwar period International Harvester had concentrated on developing small tractors to the detriment of their big tractor line. Belatedly realizing that big farmers were willing to spend big money on more powerful tractors, Harvester engineers rushed their most powerful row-crop tractor ever, the 560, into production in 1958. The basis for the new tractor was a 60-horsepower six-cylinder engine mated to a Farmall Super MTA drivetrain. The Super MTA drivetrain was itself designed in 1938 for the 33-horsepower Farmall

M. The 560 was a disaster, causing the biggest recall in International's history. The year it was introduced was the year John Deere surpassed International to become the dominant tractor manufacturer in America, a position Deere never relinquished. Easily the biggest blunder in company history, the 560 marked the beginning of a progressive downward slide in International's fortunes that eventually culminated in the failure of the company in the 1980s.

LEFT: The 560 packed a lot of power into a compact and uncluttered frame.

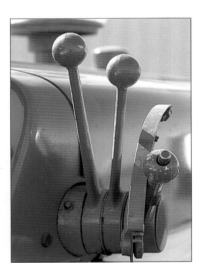

ABOVE: IH simplified operation of increasingly complicated tractors by clustering controls of similar systems in one area.

RIGHT: Gauges on the 560 were located within a couple of feet of the operator's face, a concession to comfort that was long overdue.

RIGHT: Pedal surfaces were made rough for better grip.

The two-point "Fast Hitch" was IHC's answer to the universal three-point hitch.

SPECIFICATIONS

Weight (pounds): *6020*

Transmission: *5/2-speed*

Engine (displacement, rpm): *263ci, 1800*

Horsepower (belt, drawbar): *55.46, 44.60*

Wheels and tires (front, rear): *6.50x16, 15.50x38*

Years produced: *1958–1963*

Numbers built: *66,000*

Price new: *$4090*

Owner of machine pictured: *Paul Brain*

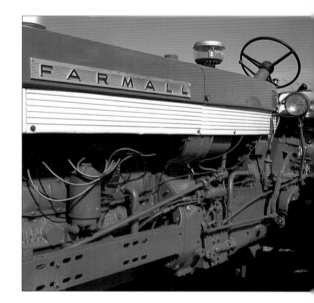

ABOVE: When not pressed to work continuously at maximum horsepower, the 560 was a good tractor. When pushed to its limits, its weak rear axle failed, earning the 560 a poor reputation at a time when the competition was building some extraordinary tractors.

*International considered the 2+2 concept to be
a major part of its future tractor line. No other
manufacturers shared this vision.*

*BELOW: International's Control
Center cab, seen here on a Model
986, was a major improvement
in operator comfort.*

*ABOVE: The Model 1066 offered
an engine developing 115
horsepower with either a gear
drive or hydrostatic transmission.*

The 1970s International started the decade with a completely new line of tractors. At the bottom of the horsepower list were three utility tractors, the 354, 454, and 574 of 32, 40, and 52 horsepower. The 574 was also built as a row-crop. In the middle of the field were the new 66 series of tractors, the 766 (79hp gas, gear drive), 966 (95hp diesel, gear or hydrostatic drive), 1066 (115hp diesel, gear or hydrostatic drive). The big tractors all had diesel engines and gear drive. They were the 1466 (133hp), 1468 (145hp V-8), and the 4166 (150hp 4WD). Standard-tread versions of the 966, 1066, and 1466 row-crop tractors followed.

An event occurred in 1973 that, though nearly unnoticed at the time, is destined to be noted in every history of the International Harvester Company. After 50 years, the name Farmall was dropped. Modern farming practices had practically eliminated the row-crop tractor. The distinction between row-crop and standard-tread tractors had become so insignificant that it no longer made sense to distinguish between the two.

In 1976 International unveiled its "86" series of tractors. These tractors offered small mechanical changes, but big improvements in operator comfort. A new cab (IHC called it a Control Center) incorporated every modern comfort feature. It looked and felt more like the cab of a new truck than like a farm tractor. The 86 tractors, ranging from the 85hp 886 to the 160hp 1586, were designed with the cab as an integral part of the tractor. These tractors came standard as two-wheel-drive with axles extended for dual rear wheels or with front-wheel-assist. A four-wheel-drive, four-wheel-steer tractor, the 4186 with 157 horsepower, and three articulated four-wheel-drive tractors topped out the IHC line. The 230hp 4386, 300hp 4586, and the 350hp 4786 were assembled for International by Steiger.

The last new development of the 1970s was the unveiling of the

A priority was replacing the troublesome 460, 560, and 660 tractors with more powerful, more durable models. The 72 PTO horsepower 706 superseded the 560, and the 806, with 94 horsepower, set a new record for row-crop power at IHC. The 37hp 404 replaced the 340 and the 46hp 504 replaced the 460. These new tractors formed the core of the IHC offering through 1968. In 1965 the 1206 Turbo, a beefed-up and turbocharged version of the 806, became IHC's first row-crop tractor with over 100 horsepower. New attention was paid to the Utility tractor category with the 36hp 424.

In 1967 the "56" series was introduced. The 756, 856, and 1256 Turbo were "06" models with improved operator comfort and safety features. A new addition was the 1456 Turbo model, with 131 horsepower. The big news of the late 1960s was the hydrostatic transmission. Introduced on the 66hp Model 656, this transmission allowed a continuously variable gear ratio between 9mph in reverse and a 20mph forward. A second range provided more wheel torque and 4mph reverse to 8mph forward. In 1969, the 113hp 826 and 112hp 1026 joined the line.

ABOVE: Big tractors like this 145-horsepower 1486 are usually seen with the dual rear wheels or front-wheel-assist needed to put all that power on the ground.

130hp 3388 and the 150hp 3588 2+2 tractors. The 2+2 was a new concept in farm tractors, an articulated four-wheel-drive row-crop machine. The engine was located in the front half, with a standard Control Center cab mounted on the rear. The tractors used engines and final drives taken from existing models to save development cost.

The 1980s The company made a major effort to design a superior transmission in 1980. The result was the Vari-Range, which combined a hydrostatic transmission and gear transmission. The Vari-Range combined the efficiency of a gear transmission with the infinitely variable speed of a hydrostatic transmission.

A 170hp Model 3788 2+2 joined the two smaller models in 1980, and plans were in the works for larger models. International considered the 2+2 concept to be a major

part of its future tractor line. No other manufacturers shared this vision, as no one else offered an articulated tractor with sliding axle tread adjustment. In 1984, before the concept could be fully developed, the IH Farm Equipment division was sold to Tenneco Corporation. International became a part of Tenneco's J.I. Case subsidiary. On May 14th, 1985, the last International tractor came off the Farmall Plant assembly line. Though there are no longer any International tractors, the new transmissions live on behind Case engines in Case-IH tractors.

LEFT: International had big plans for the 2+2 at the time of the Case buyout. The 60 series tractors were to be replaced by exciting new models with advanced transmissions and improved comfort and efficiency.

CHAPTER SEVEN

THE AMERICAN TRACTOR

★

JOHN DEERE

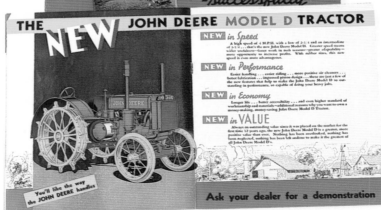

The world's premier manufacturer of farm tractors came late and rather reluctantly to the industry. Some western branch houses had begun marketing the Gas Traction Engine Company's Big Four 30 in 1910, but this behemoth was only useful in the western wheatland territories. Company director Willard Velie urged the company to investigate the possibilities for a general purpose tractor and, by a small majority, prevailed over strong opposition. John Deere's first experiments were begun in 1912 by C.H. Melvin. Melvin's designs came to naught, but Joseph Dain continued experimenting for the company until 1918, when his 25hp, three-wheeled, all-wheel-drive tractor was approved for production. Deere built 100 of these units over two years.

Tractor experiments were only progressing slowly and Deere management's commitment to the tractor trade was weak. Nevertheless, dealers and branch houses were clamoring for a tractor to sell.

ABOVE: The first commercially successful John Deere tractor was the Model D, introduced in 1923. It was not an advanced design even then, but it was so simple and durable it remained in production for 30 years.

Deere a line of gas engines as well as the Waterloo Boy tractor, a dependable, though not particularly advanced, 25hp machine. The Waterloo Boy could be sold for much less than John Deere's tractor, and if the tractor venture failed, the well-proven market in gas engines would salvage the expenditure on the purchase.

John Deere placated its dealers with the Waterloo Boy, but continued with experiments on its own tractor. In 1923, John Deere unveiled its Model D, the progenitor of a line of two-cylinder tractors destined to endure for nearly 40 years. Outwardly the Model D appeared crude. It had a huge, slow-turning, two-cylinder engine with an exposed, spoked flywheel design—features that were rapidly being abandoned by other tractor manufacturers. But the D was advanced in many ways. Like other new designs, its transmission and engine cases were incorporated into a unit frame, giving it strength and rigidity while keeping weight to less than 5000 pounds. Its engine

Waterloo Boy In 1918, in order to get a small tractor into its dealer's hands quickly with a minimum of financial risk, John Deere bought the Waterloo Gas Engine Company. The Waterloo purchase brought John produced 41 brake horsepower and was designed to burn low-cost distillate fuel efficiently. The D's 2-speed transmission and final drive unit was a model of simplicity and efficiency and was sealed from dust and dirt.

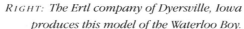

RIGHT: The Ertl company of Dyersville, Iowa produces this model of the Waterloo Boy.

LEFT: The Waterloo purchase included an impressive plant as well as two tractor designs to build in it. Waterloo Boy tractors kept John Deere in the business until the Model D was ready five years later.

ABOVE: When developing a tractor of their own took longer than anticipated, John Deere bought the Waterloo Gas Engine Company in 1918 and immediately had a tractor to sell—the Waterloo Boy.

Though experiments with a row-crop tractor soon followed the introduction of the Model D, it was the only tractor in the John Deere line until a few Model C row-crops were released in 1927. In 1928 the Model GP, Deere's first production row-crop tractor was unveiled. At this time in the development of the row-crop tractor, the narrow front wheel/adjustable rear wheel configuration that eventually became the norm was not yet assumed to be the best design. The Farmall was designed to cultivate two or four rows of crops, with the narrow front wheels running between two rows. John Deere's GP had high ground clearance like the

1923–1953
John Deere D

Prior to the introduction of the Model D, John Deere was barely in the tractor business. Deere experimented with many machines, but continued to build the woefully obsolete Waterloo Boy. At the insistence of Leon Clausen, who later would lead J.I. Case to its success in the tractor field, Deere developed and built a more modern version of the Waterloo Boy, the Model D. The D was an instant success, selling 1000 tractors the first year. With a two-cylinder engine and 2-speed transmission, the Model D was strong and simple, containing fewer parts than any other tractor of its size. The John Deere D is the machine that established John Deere as a serious tractor manufacturer and for 30 years was the touchstone against which all John Deere tractors were measured.

The John Deere D offered the ultimate in simplicity. The D had fewer parts than any tractor of its size, but it was, nevertheless, a powerful and rugged tractor that would do good work for years with minimal care.

LEFT: With its low seat and plenty of room to stand when sitting became tiresome, the D had one of the more comfortable operator's stations.

RIGHT: The engine was started and warmed up on gasoline. The operator's manual said the switch to kerosene could be made when the top of the radiator was, "too hot to hold your hand on."

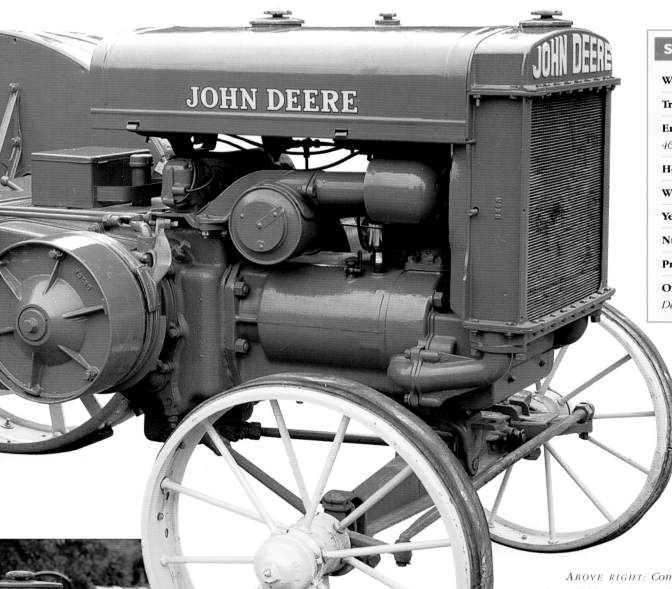

SPECIFICATIONS

Weight (pounds): *4403*

Transmission: *2-speed*

Engine (displacement, rpm):
464.6ci, 900

Horsepower (belt, drawbar): *27, 15*

Wheels and tires (front, rear): *28x5, 46x12*

Years produced: *1923–1953*

Numbers built: *161,270*

Price new: *$1125*

Owners of machine pictured:
David and Paul Palmer

ABOVE RIGHT: Compact, dependable, and powerful, for 30 years the D was the tractor of choice on many medium-sized American farms.

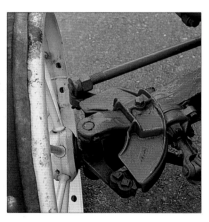

RIGHT: The exposed steering gear was subject to premature wear. Like most parts on the D, replacement gears were inexpensive and easy to fit.

*"The Famous John Deere Farm Tractor
—Its Best Salesmen Are The Men Who Use It."*

– *Newspaper advertisement for the Model D, 1928*

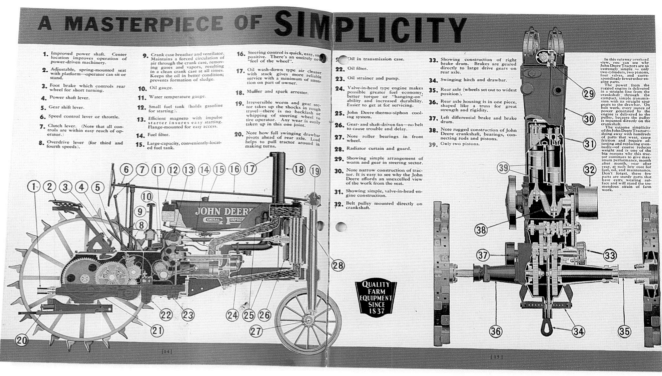

RIGHT AND BELOW: John Deere tractors truly were masterpieces of simplicity. They produced comparable horsepower using fewer moving parts than other tractors in their weight class. The two-cylinder design lasted for 37 years.

LEFT: This 1935 Model B is a classic American tractor. Its row-crop design with steel wheels was the most popular configuration sold in the mid-1930s.

Farmall, but its front wheels were widely spaced, allowing each front wheel to run in the same row as its corresponding rear wheel. John Deere felt this design held several significant advantages; the GP left only two wheel tracks in the field instead of the Farmall's three, and it kept Deere from infringing on patent rights held by IHC.

with row-crop tractors. The Model A was superior to every other tractor of its type. It combined the Model D's simplicity with the features the engineers had learned were important; high crop clearance with narrow front wheels, adjustable rear tread, a comfortable operator's platform,

The 1930s John Deere's first row-crop tractor was not nearly as successful as the Model D and Deere never seemed happy with it. Development continued through 1934, the year John Deere's second important tractor was unveiled—the Model A. John Deere engineers learned their lessons well during the many years spent experimenting

and excellent visibility. The Model A's rear wheels were mounted on splined axles and could be adjusted for a wide variety of crops. Its two-cylinder engine produced 24 horsepower, plenty for a medium-sized thresher or shredder. With a 4-speed transmission and 18 drawbar horsepower it could handle a two-bottom plow in almost any conditions.

ABOVE: AR and BR tractors always received updates after the better selling row-crop models.

ABOVE: The Model D was in its heyday in the mid-1930s. Until the Model G came out in 1938, it was the most powerful John Deere tractor available

BELOW: The BR was the standard-tread version of the Model B. The row-crop tractor outsold the standard-tread many times over.

RIGHT: The John Deere name was cast into the radiators of over 300,000 tractors in the 1930s.

The Model A could handle a four-row cultivator and lift it nearly effortlessly at the end of the rows with an hydraulic lift. The hood tapered back from the radiator to only a few inches wide at the rear, allowing excellent visibility of the rows being cultivated. It took 22 years, but Willard Velie finally had his general purpose tractor.

Immediately on the heels of the Model A came a scaled-down version called the Model B. The B had all the features of the Model A in a 16 brake horsepower, 2763-pound tractor. Both Models A and B came in a variety of styles. There were an abundance of front axle, tire, and wheel configurations that adapted the tractors to almost any crop or need.

In 1935 standard-tread versions of both Models B and A were introduced. These models, called the AR and BR, used Model A and B engines and other components. John Deere then had a line of standard-

1939
John Deere A

As with the standard-tread Model D, John Deere's row-crop Model A introduced little that was new to the farm tractor itself. Success was all in the execution. The Model A did its job with fewer parts and greater simplicity than any other row-crop tractor on the market. With a two-cylinder engine specially designed to burn low-cost fuel, the Model A set records for fuel economy. Its chassis design was similar to other row-crop tractors, but it was just a little narrower, visibility was a bit better, and it was easier and more comfortable to operate. Its hydraulic power lift was a little smoother and easier to operate than the competition's mechanical lifts. With fewer parts to break and wear out, the Model A cost less to maintain. Though it offered little in the way of innovation, the John Deere A built a sterling reputation on simplicity, dependability, economy, and the feature farmers valued most—value.

ABOVE: The hydraulic lift was a popular option. Cultivators and other equipment could be lifted by simply stepping on a pedal or pushing a lever. The hydraulic lift introduced on the Model A in 1934 saved countless repetitions of the back-breaking work of lifting cultivators by hand.

LEFT: Opening the compression release made it easier to crank the tractor to start it.

SPECIFICATIONS

Weight (pounds): *4059*

Transmission: *4-speed*

Engine (displacement, rpm): *309ci, 975*

Horsepower (belt, drawbar): *23.52, 16.22*

Wheels and tires (front, rear): *24x4, 50x6*

Years produced: *1934–1952*

Numbers built: *209,000*

Price new: *$850*

Owner of machine pictured:
Gene Gabbert

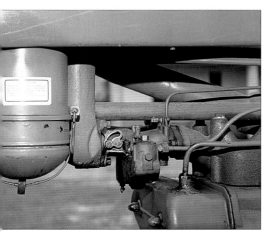

BELOW: Open access to vital parts made routine maintenance on the Model A easy.

The Model A's art deco style was developed by famed industrial designer Henry Dreyfuss. Even with a wide front axle the tractor is a trim-looking machine. The majority of John Deere As were shipped with a tricycle-type front end. This wide front chassis, officially called the AW, was popular in hilly areas.

ABOVE: Like all John Deeres, the Model A engine was simpler and had fewer moving parts than any tractor of similar ability. The A was designed from the beginning to operate on distillate fuel.

"Figuring Your Costs or Working In The Field
—You'll Be Mighty Glad You Bought a John Deere."

– Magazine advertisement for the two-cylinder range, December 1939

tread tractors ranging from 16 to 41 horsepower. The Model GP, which had been filling the position as a smaller standard tractor after the A was introduced, was dropped when the AR and BR debuted.

For many farmers even the Model B was too much tractor. Deere set out down two separate paths to fulfil the needs of the small vegetable and tobacco farmers. One path, initiated by Deere's Wagon Works, led them to the Model L tractor. Introduced in 1937, with a vertical two-cylinder engine and a pipe frame, the L was completely different than anything

ABOVE: The AOS was one of two orchard versions of the Model A. The AOS provided additional shielding and a more streamlined shape to push through low-hanging branches better. It was expensive to build and only remained in production from 1936 through 1940.

Deere had ever produced. With nine horsepower, weighing only 1570 pounds, and at a price of only $658, it fit the needs of the very small farmer well.

The other path was followed by Deere's tractor works in Waterloo, Iowa and led, in 1939, to the Model H. The H was outwardly a scaled-down Model B. Deere's attempt to reduce the Model B was not as easy or successful as shrinking the A. The H was a good enough machine, but it shared few parts with the other tractors, making it relatively expensive. At $855 for only 14 horsepower when the improved 20hp B cost only about $1000, the H offered expensive horsepower indeed.

John Deere also perceived a need at the other end of the power

spectrum, and offered the Model G in 1938. The G was a row-crop tractor like the Models A and B, but with a 413ci engine and 36 horsepower it was substantially larger.

In 1939 the Model B received more power from a larger engine. At the same time, both the A and the B were clothed in stylish new sheet metal fashioned by industrial designer Henry Dreyfuss. The Model H was given this distinctive look when it came along in 1939, but no other models received it until 1941. Dreyfuss' styling came to the Model D in 1939, along with optional electric starting and lights. Styled sheet metal was still not offered on the AR or BR.

ABOVE: John Deere advertising frequently included children. It was designed not only to appeal to children, but to parents who wanted to keep their offspring on the farm. The clear message is Jr. is happy driving a John Deere.

ABOVE: *Compare this Model AO with the AOS on the previous page. When farmers discovered that in most instances all that sheet metal was unnecessary as well as in the way, it got left out behind the shed.*

LEFT: *Once again, John Deere advertising appeals to the family farmer. Inside it stresses the benefits of a John Deere tractor: simplicity, fuel efficiency, and speed to get him back home to his family sooner.*

The 1940s John Deere entered the war decade with a tractor for nearly every need. With sizes from eight to 36 horsepower, its line of row-crop tractors was the most complete in the industry. The standard-tread line lacked only a four-plow tractor. More powerful tractors, electric starting, diesel engines, and increased use of hydraulic power would characterize tractor development in the 1940s, and John Deere engineers were ready for the challenge.

The first new development of the decade was to increase the size of the Model A's engine from 309ci to 321ci, providing a little more power. Electric starting was offered on the A and B for the first time. The next year a larger version of the Model L, the LA, was added to the line. The LA very similar to the L, but could boast of 13 horsepower. The next order of business was to give the Models A and B 6-speed transmissions.

In late 1941 the Model G was wrapped in sheet metal styled to match that of the A, B, and H. The styled G was called the GM until the end of the war, at which time its designation reverted to simply G. For the next five years the only major change in the row-crop line was the addition of Powr-Trol in 1945. Powr-Trol was John Deere's implementation of a hydraulic power lift. The hydraulic lift introduced with the Model A in 1934 consisted of a rockshaft that was either up or

RIGHT: Roll-O-Matic was introduced in 1947 to help take the jolts out of steering.

down – not much different, really, from the mechanical lifts on other tractors. But Powr-Trol allowed precise control of the degree of lift, plus it offered an hydraulic outlet that allowed it to power remote hydraulic lift cylinders.

The year 1947 was a big one for John Deere. The Models A and B received significant improvements. The electric starting system, which previously had been a tacked-on afterthought, was integrated into the tractors and made standard equipment. A new seat provided a conveniently located mounting box for the battery as well as increased comfort for the driver. A stamped steel frame took the place of the old cast angle frame. The B received a newly designed transmission, still with six speeds, as well as an even larger engine and a small power increase. Both the A and B models offered high-compression engines optimized for gasoline fuel. Roll-O-Matic, a knee-action joint between the front wheels that increased driver comfort and safety, was available on A, B, and G.

The postwar Model G was little changed from the GM. It received the new seat and battery box unit and electric starting was made standard equipment, but the starting system was the same tacked-on unit offered previously. A gasoline fuel option was not offered.

The All-New M The first big event after the war was the announcement of the Model M. The M was an all-new tractor with a vertical 101ci two-cylinder engine with its crankshaft mounted longitudinally. The engine and transmission were joined by a cast iron tube, and together they formed the tractor's frame.

The Model M had 20 horsepower, a 4-speed transmission, and came with electric starting as standard equipment. The influence of the Ford tractor was obvious in the Quik-Tatch hitch system with hydraulic lift and specially designed mounted implements. John Deere expected big things from the M. The company dropped the H, L, and LA and allowed the M to fill their positions at the small end of the line. The BR and BO orchard models were also eliminated and the Model M was the closest thing Deere had to fill those gaps. With five tractor models to replace, the M needed to be a very versatile machine. John Deere sold it as the M with an adjustable wide front axle, as the MT with a tricycle chassis, and as the MC when equipped with crawler tracks. An industrial version, the MI, with lower, shorter chassis and fixed wide front axle was also available.

The standard-tread line was given much less attention. The Model D continued much as it was in 1939. Powr-Trol

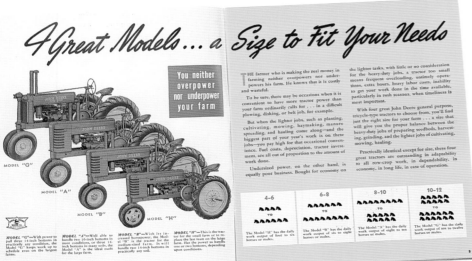

ABOVE: "You neither overpower nor underpower your farm." In 1940 John Deere made it possible to take this advice with these four row-crop tractors with 14 to 37 horsepower. For the really small farm there was the tiny Model L.

The R had all the size, convenience, and power large acreage
farmers were clamoring for, but it excelled in yet another area.
It proved to be the most fuel-efficient tractor ever built.

BELOW: The Model R carried a colorful "Leaping Deere" logo for the first time.

RIGHT: The Lindeman Company of Yakima, Washington, began converting John Deere tractors to crawler chassis in the 1920s. The BO model converted to tracks was so successful, John Deere bought the company in 1946.

ABOVE: John Deere's first diesel, the Model R, was an immediate success and spawned a series of two-cylinder diesel tractors.

was made available on the Model AR, but it was otherwise unchanged and the BR was dropped entirely.

No further changes were made until January 1949, when John Deere dealers began receiving the Model R, a radical new standard-tread tractor with an impressive list of John Deere firsts. At 51 maximum horsepower, the R was the first five-plow tractor John Deere had ever produced. It had live power take-off and a live Powr-Trol hydraulic system. Its most impressive feature was the 416ci two-cylinder engine. It was a diesel. The R had all the size, convenience, and power large acreage farmers were clamoring for, but it excelled in yet another area. It proved to be the most fuel-efficient tractor ever built. At a price of $3800 farmers bought them as fast as John Deere could make them.

Almost overshadowed by the introduction of the Model R was the modernization of the old AR. The AR received a 6-speed transmission along with the more powerful gasoline and distillate engines first offered in the Model A row-crop in 1947. It also received sleek new sheet metal that matched the styling of the Model R.

The 1950s The new decade began with a new 6-speed transmission for the Model A, introduced in late 1949 for the 1950 model year. Two tractors, the models AH and GH were added to the line. These were A and G row-crops modified to provide 48 inches of crop clearance under the axles. Few changes were made in the line

"It's Powerful Good News!...The New John Deere 420 Fleet"

– Advertisement for the 420 model range, mid-1950s

ABOVE: The row-crop 60 took the place of the Model A and was the most popular and versatile model of the series. It was available in row-crop, orchard, high-crop, and standard-tread chassis.

until the unveiling of a completely new series of tractors began in 1952.

The rework of the model line that began in 1952 was much more cosmetic than that of 1947. The first tractors affected were the Models A and B. With the addition of live power take-off, a new carburetor, and other engine changes, and new styling that matched that of the R and AR, the Model A became the 60. The Model B received similar changes and became the 50. The John Deere tractor that started it all, the venerable Model D, was dropped from the line.

In 1953 the Models G and M received similar treatment to become the 70 and the 40. The R continued unchanged. Gasoline fuel was available on the 70, and LP fuel became an option on the 60 and 70. John Deere finally got a three-point hitch, at least for the Models 50, 60, and 70.

In 1954 John Deere unveiled the diesel-powered Model 70. For decades John Deere had based its reputation on efficient use of fuel, and with the diesel 70 it set a mark that wasn't matched for years. The 70 was available as a standard-tread or row-crop tractor, making it John Deere's first diesel row-crop.

The following year saw the Model R replaced by the 80. The 80 offered a 6-speed transmission and a more powerful engine in a tractor that was otherwise quite similar to the R.

ABOVE: The Model 720 was little changed from the record-setting Model 70. Live power take-off, a three-point hitch with draft control, and standard power steering were the major new features.

BELOW: While other makers were introducing shift-on-the-go transmissions, John Deere owners had to be content with a gated 6-speed. They contended that with the two-cylinder's lugging power, on-the-go downshifts were unnecessary.

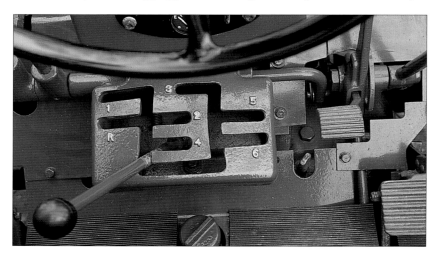

In 1956 a minor rearrangement of the line began, with the introduction of a more powerful version of the model 40, known as the 420. The 40 was continued, little changed, as the 320. The rearrangement continued in 1956, with the 50, 60, 70, and 80 becoming the 520, 620, 720, and 820. More power was available across the board and a draft controlling three-point hitch was available on all but the 820. The earlier sheet metal remained, but a bold yellow strip was painted on the hood sides.

The last tractors of the two-cylinder line were introduced in 1958. This was the 30 series of tractors. The 330 through 730 tractors were the same as corresponding 20 series tractors except for a more comfortable steering wheel position and revised sheet metal. The 830 differed from the 820 only in that it had the new sheet metal.

BELOW: The little 320 was a prettied-up version of the popular Model 40.

John Deere offered a wide variety of options on the 420/430 tractors, including a broad assortment of wheel and axle combinations, a crawler chassis (the only crawler in the John Deere line), and three different fuel options. In 1959 Deere expanded the options even more with the Model 435, a 430 equipped with a General Motors two-cylinder diesel engine.

Six-Cylinder Power The 1950s ended on a dramatic note. At the 1959 introduction of the 30 series tractors, the company demonstrated a huge, articulated four-wheel-drive tractor of 215 horsepower. The massive 8010, the first 6-cylinder tractor offered by John Deere, was in fact purchased from Wagner, a specialty tractor manufacturer. It gave the first hint

ABOVE: Cane farming is notoriously hard on tractors. The John Deere 70 high-crop stood up well to harsh climate and rough handling.

1955
John Deere 70 Diesel

Through the late 1940s and early 1950s John Deere watched from the sidelines as one by one Oliver, Massey-Harris, Farmall, and Minneapolis-Moline brought out diesel-powered row-crop tractors. With experience gained on the Model R, by 1953 John Deere had the design and manufacturing experience to produce a superior diesel row-crop. When it was tested at the University of Nebraska in 1954, the results showed John Deere had triumphed again. With a rated horsepower of 42.8 and a maximum of 48.29, the 70 fit in the top power echelon of row-crop tractors. But it was with its use of fuel that the 70 was truly superior. With a fuel economy rating of 17.74 horsepower-hours per gallon, the Model 70 set a fuel economy record that stood for over a decade. The John Deere 70 paid the row-crop farmer back for his investment faster than any other tractor and set the stage for a revolution in farm tractor engines. Though none could match its economy, every year following the success of the Model 70 saw the introduction of another diesel row-crop tractor. Today, diesel is the rule rather than the exception in row-crop farm power.

Power Steering

LEFT: Power steering was a much needed and appreciated option on the 70.

RIGHT: The serial number 7033054 reveals this to be a 1955 Model 70.

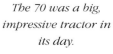

The 70 was a big, impressive tractor in its day.

A four-cylinder gasoline motor was used to start the big two-cylinder diesel. Both engines are housed under the narrow hood. Access to the starting motor is through the door just behind the "John Deere" logo.

SPECIFICATIONS

Weight (pounds): *6035*

Transmission: *6-speed*

Engine (displacement, rpm): *379.5ci, 975*

Horsepower (belt, drawbar): *42.80, 33.16*

Wheels and tires (front, rear):
6.00x16, 13.00x38

Years produced: *1953–1956*

Numbers built: *17,000*

Price new: *$2692*

Owner of machine pictured:
Greg Kokemiller

LEFT: Though the fuel is heavily filtered, water and large debris fall out of the fuel in this glass sediment bowl.

LEFT: In spite of its size, its narrow hood and easy maneuverability made the 70 an excellent cultivating tractor.

The fluted grille is designed to be easily cleaned.

LEFT: A comprehensive set of engine gauges was standard on the 70. The large dial in the center monitored engine rpm as well as recording total hours of engine operation. Controls for the starting engine are to the right of the main panel.

that John Deere might be willing to stray from the two-cylinder engine design it had championed for nearly 40 years.

John Deere had entered the decade with six different tractor models that were available with only one or two fuel options. As farmers sought greater efficiency and profits, distillate, Deere's mainstay fuel, fell out of favor. Deere responded by offering more tractors with LP and diesel engines. John Deere also managed to stay near the industry leader Ford in the use of hydraulics. Deere avoided the trap of insisting on its own hitching system and adopted Ford's three-point hitch early on. By the end of the decade John Deere was the industry's sales leader.

The 1960s As the 1960s began, farmers were happily buying John Deere's 30 series tractors in near record numbers. The industry barely stirred when it was announced that the 1961 tractors were to debut in Dallas, Texas on August 30. The first hint that this would be something special was the planned size of the event. Over 6000 people were to attend and the entire city of Dallas was involved. When the wraps were

STEP UP to farming ease and convenience

STEP UP to JOHN DEERE

ABOVE: *John Deere continued to advertise two-cylinder tractors as if the supply would never end, knowing all the while that the days of the "Johnny Popper" were numbered.*

lifted on the new machines, the 1961 tractors were revealed to an utterly stunned industry. John Deere had scrapped its entire tractor line. In its place were machines that virtually redefined the farm tractor. The only thing the same was the green and yellow paint.

John Deere called its new line the "New Generation" and a more appropriate name can hardly be imagined. There were six series of tractors with power ranging from 36 to 215 horsepower.

The smallest tractor, the 36hp 1010, weighed 5754 pounds. It had a four-cylinder gasoline or diesel engine and a 5-speed transmission. It was available as an orchard and grove tractor, a crawler, utility, or row-crop with several wheel and axle options. The 2010 offered similar equipment in a 47hp tractor with an 8-speed Syncro-Range transmission. Deere's Syncro-Range allowed shifting between two gears in each of four ranges without bringing the tractor to a stop.

The 59hp 3010 was available as an orchard, standard-tread, wide-front row-crop, and narrow-front row-crop chassis. It incorporated all of Deere's newest innovations, including the advanced closed center hydraulic system. Closed center hydraulics used a single large pump to provide hydraulic power for the implement lift, the power steering, power brakes, and three remote hydraulic circuits. All New Generation tractors placed a new emphasis on operator comfort. Henry Dreyfuss' team started with a clean slate and redesigned the entire operator's area to optimize safety and comfort and minimize fatigue.

The 4010 was expected to be the biggest seller. Farmers had been buying large row-crop tractors in greater numbers and the 4010 was the largest row-crop Deere had ever offered. Though it weighed

LEFT: *The last of the two-cylinder diesels was the Model 435. It was a response to Massey-Ferguson's popular small diesel utility tractor.*

BELOW: Some owners argue that the 4020 was the best John Deere tractor ever.

RIGHT: John Deere's success with the 3020 and 4020 did not extend down to its smaller tractors.

BELOW: With 108 horsepower, the 5010 was the brute of the John Deere line.

Here's where modern engineering design really shows

1. Heavy-duty 4-cylinder variable speed Diesel engine.
2. Front-mounted fuel tank holds a full day's supply for average work-loads ; front location reduces fuel evaporation and fire hazard.
3. Direct fuel injection ensures fast starting and good torque reserve for increased pulling power.
4. Crankshaft main bearings overlap more than 25% of the connecting rod journals for extra rigidity.
5. Exclusive variable-displacement pump supplies full hydraulic power on demand.
6. Eight uniformly spaced forward speeds ; four reverse speeds.
7. Oil-cooled hydraulic brakes with hydraulic interlock never need adjusting.
8. Fast, effortless power steering available as an option ; mechanical linkage ensures safe steering when engine is not running.
9. 1000 rpm 'live' mid-point PTO operates independently of rear PTO.
10. Two rear hydraulic outlets for single or double-acting remote cylinders ; mid-point outlets also available.
11. Easily adjusted universal 3-point hitch handles Category 1 and 2 equipment.
12. Lower-link sensing responds instantly, accurately, to change in draught.
13. Two 'live' rear PTO options : 540 rpm and dual-speed 540/1000 rpm.
14. Multi-position, steel-disc rear wheels provide various tread adjustments.
15. Planetary final drives distribute gear load, lengthen gear and bearing life.
16. De luxe seat is adjustable for your height and weight for a more comfortable ride.

Cutaway view of 4-cylinder 2020. In certain details, explained elsewhere in this brochure, other power sizes are slightly different.

joined the line in 1962.

The 5010 shared the features of a 4010 standard scaled up to handle its 121 horsepower. It was available only as a standard tractor. It was the industry's first two-wheel-drive tractor with over 100 horsepower and the first to use the heavy duty category 3 three-point hitch.

The New Line-Up By 1963 the New Generation line was fully in place. It consisted of six sizes of tractors in 19 versions. The 1010, 2010, and 3010 provided the line's utility tractors. The 1010, 2010, 3010, and 4010 were available as row-crops, and 3010, 4010, and 5010 comprised the standard-tread line-up. The 8010 provided a four-wheel-drive articulated tractor. A single row-crop tractor was available on the 1010 platform. Grove and orchard tractors came on the 1010 and 3010 chassis, while the company's high-crop tractors were built on 2010 and 4010 chassis.

less than the 730, the 4010 gave over 50 per cent more horsepower for a cost difference of only about 18 per cent, making the 4010 tremendous value. The 4010 had all the features of the 3010, but was powered by a six-cylinder version of the 3010's four-cylinder engine. Gasoline, diesel, and LP were the fuel options.

The 215hp 8010 was continued from 1960, leaving a huge power gap between it and the 4010. A tractor was planned to fill this gap and it finally

L A N D M A R K   M A C H I N E S

1963
John Deere 4010

I t would be a monumental and tricky task for John Deere marketing executives to overcome the decades of declaring two cylinders to be all any farm tractor needed. For nearly 40 years John Deere owners had accepted the two-cylinder design as the only sensible way to build a tractor, and John Deere customer loyalty was legendary. It would take an extraordinary machine to convince owners of the old Johnny Poppers that a tractor with four or six cylinders was worthy of the John Deere name. The 4010 did just that. The

tractors that came later, particularly the 4020, had more dramatically advanced features, but it was the 4010 that convinced farmers that John Deere had not lost its mind or abandoned its customers. With superior comfort and ease of use, economy and efficiency that nearly matched the two-cylinder tractors, and John Deere's famous customer support, the 4010 paved the way for a new generation of farm tractors from John Deere.

John Deere rethought everything. The fuel tank was located in front of the radiator, a position which keeps the fuel cooler so making refueling safer.

LEFT: Engine, transmission, hydraulic, and PTO controls are all at the operator's fingertips. Engine rpm, charging circuit current, and fuel level are displayed on gauges, while oil pressure and water temperature have only warning lights.

LEFT: A 550rpm and 1000rpm power take-off were available.

ABOVE: The 4010 fuel system was simple and easy to maintain.

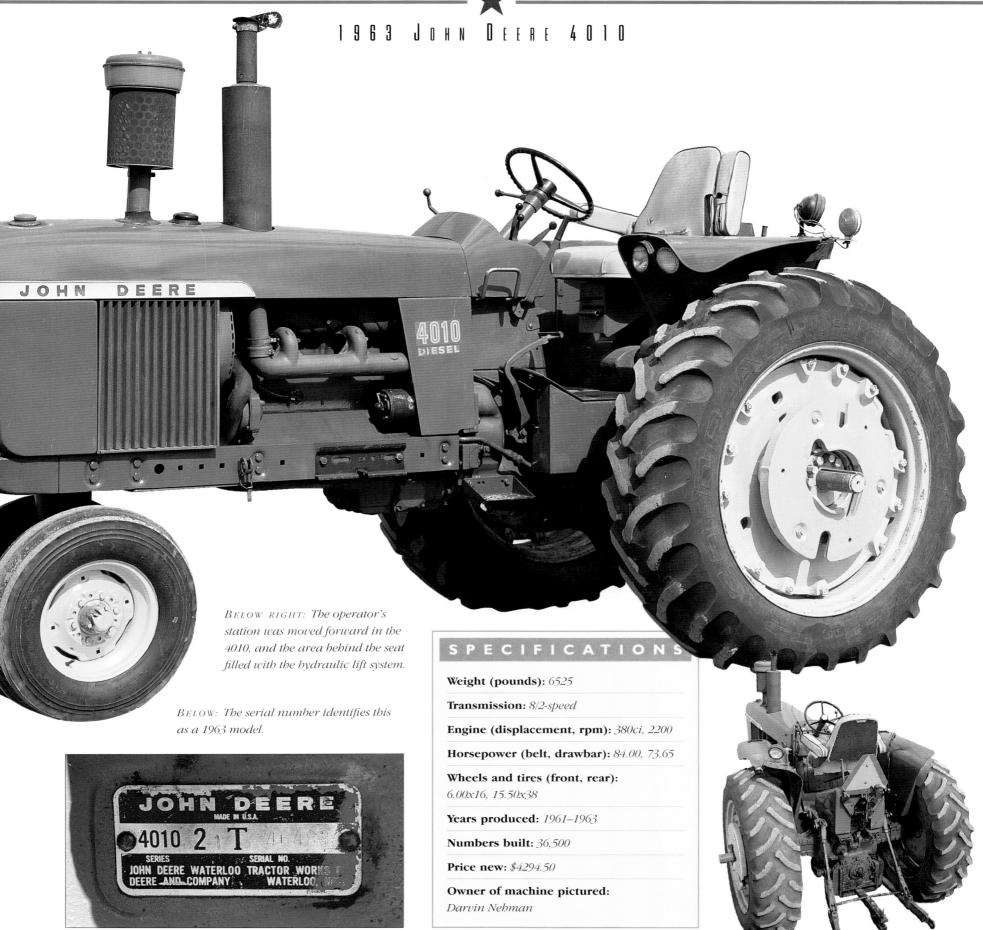

BELOW RIGHT: The operator's station was moved forward in the 4010, and the area behind the seat filled with the hydraulic lift system.

BELOW: The serial number identifies this as a 1963 model.

SPECIFICATIONS

Weight (pounds): *6525*

Transmission: *8/2-speed*

Engine (displacement, rpm): *380ci, 2200*

Horsepower (belt, drawbar): *84.00, 73.65*

Wheels and tires (front, rear):
6.00x16, 15.50x38

Years produced: *1961–1963*

Numbers built: *36,500*

Price new: *$4294.50*

Owner of machine pictured:
Darvin Nehman

"...a tractor of surprises...in its willingness to take the heavy work off your shoulders, increase your profits and part the curtain to that long-sought commodity called spare time." – *5020 sales brochure*

Eleven models—34 to 143 horsepower...

New 820
34 horse power—
tough, compact, ready to
tackle any one of a hundred
different jobs.

New 920
40 horse power—
ideal for small and medium
acreage farms, or as a
second tractor on big-scale
operations.

New 1020 Series
47 horse power—
compact, stable, versatile.
Available in Utility, Vineyard
and Orchard models.

New 1120
52 horse power
powerful all-round tractor,
ideal both for cultivating
and all kinds of utility jobs.

New 2020 Series
64 horse power—for ample
power and steady reliable
service year in, year out. Available
in Agricultural and Orchard models.

LEFT: *A common styling theme gave John Deere's New Generation line an integrated look—an important marketing consideration as it associated the legendary quality of John Deere's Waterloo models with all tractors in the line. Underneath, the vehicles were quite different, built in different factories in different countries.*

It is a tribute to the thoroughness of John Deere's engineering that totally new tractors could put in the field without recalls, second thoughts, or last minute redesigns. But time marches on, and with it marches technology. By 1963 the Power Shift transmission was ready. Power Shift was an 8-speed transmission that allowed gear changes without using the clutch. In 1963 Deere made this transmission an option on the 3010 and 4010, renaming them the 3020 and 4020. In 1965 the 5010 was given more power and renamed the 5020. Also in 1965 the Model 2510 was introduced. This was essentially a 3020 with a smaller, 54hp engine. Unlike the 3020, the 2510 was available as a high-crop.

Though their model designations were not changed, the 3020 and 4020 received power upgrades to 71 and 99 horsepower in 1966. The 2510 was upgraded to 61 horsepower and redesignated the 2520.

If the New Generation had a weak area it was at the low power end of the line. The 1010 and 2010 failed to live up to the extraordinary reputation the larger tractors were accruing. In 1966 the two smaller tractors were replaced with new tractors, the 1020 and 2020. The 1020 had a 38hp three-cylinder engine, while the 2020 had a 54hp four-cylinder engine. Both tractors were available with gas or diesel fuel options, an 8-speed transmission, and closed center hydraulics, making their catalog of features more similar to the popular larger tractors.

By the mid-1960s John Deere had become a global company with plants in Europe and North America shipping tractors all over the world. Smaller tractors had always been popular in Europe, and overseas manufacturers had developed them to a high degree. Tractors of 60 horsepower and more were popular on large America farms. Rather than

design and built two parallel lines of small tractors, John Deere concentrated domestic development efforts on its large tractors and turned to Europe to supply it with small tractors. In 1968 the 31hp 820 and 47hp 1520 were imported to fill the gaps in the small tractor line.

The M&W Gear Co. had been building a successful turbocharger system for the 4010 and 4020 for five years, but John Deere's first factory turbocharged tractor was the 4520 of 1968. This tractor was based on the 4020, but produced 122 horsepower. To help get this power to the ground, a hydrostatic front-wheel-assist was made available for it. Availability was extended to the 3020 and 4020 the next year.

John Deere ended the decade by offering a new tractor of sorts, the 4000. The list of standard features on John Deere tractors was growing wildly and, while they were useful to most, many farmers had no need for many of them. In response John Deere offered what was essentially a stripped-down 4020 with a price tag nearly $1000 less.

"Coming on strong and sound...two new John Deere Tractors featuring Sound-Gard design for the '70s."

— Sales brochure for the Models 4230 and 4430

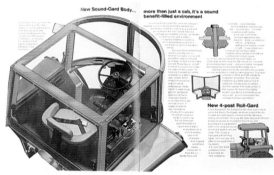

RIGHT: Ertl's toy model of the 5020.

The 1970s The landscape was changing in the tractor industry. Utility tractors had grown to over 50 horsepower and supplanted the small row-crop tractor. The trend was toward greater horsepower in all weights and types of tractors, with imported models proliferating in the under-50 horsepower area. John Deere's model line changed to reflect the new trend. Its small tractors were all utility types, with row-crop tractors filling

the mid-size line and articulated four-wheel-drive tractors in the large tractor line. With so many sizes of tractors offered and a myriad of options, it had become impossible to change the entire line in one year. The turbocharged tractors were reshuffled at the end of 1970, with the addition of the 115hp 4320 and the upgrading of the 4520, renamed 4620, to 135 horsepower.

The small end of the 20 series tractor line got its update to 30 series stature in the first years of the new decade. The 2020 was replaced with the more powerful 60hp 2030 in 1971. The 31hp 820 was upgraded to 35 horsepower and renamed the 830 in 1973. The 1520 was replaced by the 1530, a similar 45-horsepower tractor built in John Deere's Mannheim, Germany plant.

In 1972 the mid-size line was given a complete overhaul. The 3020, 4000, 4020, 4320, 4620, 5020 were all dropped from the line. Four new tractors were available, the 81hp 4030, the 100hp 4230, the 126hp 4430, and the 151hp 4630. The three largest tractors used the same engine in normally aspirated, turbocharged, and turbocharged and intercooled form. These tractors received new styling in the form of a reshaped hood. The 5020 was replaced by the 6030, with 175 horsepower. As Deere's mid-sized line grew, it became necessary to add a new tractor in the 70hp area, the 2630.

Perhaps the most important new feature was the Sound-Gard cab, more of a body really, that was available on all mid-range tractors. The cab sealed the driver away from the elements in an office-like environment surrounded by curved glass. The unit was pressurized to keep out dust, with heat, air conditioning, and a sound system optional.

While hydrostatic front-wheel-assist or mechanical front-wheel-drive was available on many of its tractors, John Deere's four-wheel-drive articulated tractor line had been neglected. The tractors had been purchase units, repainted and styled to fit John Deere's line. In 1972 John Deere announced its own

ABOVE TOP: The smallest of the U.S.-built tractors for 1977 was the 4040. With 90 horsepower it was a good size for forage harvest work.

ABOVE: The 7520 was the largest John Deere-built tractor when it was introduced in 1972, and was a tremendous success.

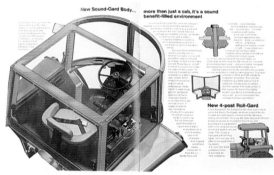

ABOVE: The Sound-Gard cab offered roll-over protection as well as insulation from deafening engine noise and a comfortable refuge from the elements.

articulated tractors, the 7020 with 146 PTO horsepower and the 175hp 7520. In 1974 these tractors were given more power and renamed the 8630 and 8430.

1976
John Deere 4430

In 1972 John Deere announced a new line of tractors that was almost as dramatic as the introduction of the New Generation of Power tractors in 1960. Previously, farm tractors had focused on the amount of work the machine could do without giving too much attention to the limits of the operator. The result was a generation of farmers with back and hearing problems. The advent of big diesel engines had made the noise problem even worse. With its Generation II tractors, the most popular of which was the 4430, Deere introduced the Sound-Gard cab. The cab had a huge wrap-around windshield, rollover protection, and optional environmental controls such as a pressurizer to keep dust out, heat, and air conditioning. The new John Deere Sound-Gard cab marked the beginning of a concern for operator health and safety that has immeasurably improved the lives of farmers.

SPECIFICATIONS

Weight (pounds): *10,735*

Transmission: *8/2-speed*

Engine (displacement, rpm): *404ci, 2200*

Horsepower (PTO): *125.8*

Wheels and tires (front, rear): *10.00x16, 18.4x38*

Years produced: *1973–1977*

Numbers built: *61,960*

Price new: *$22,575*

Owner of machine pictured: *John Schow*

BELOW: A turbocharged version of the 4020 engine powered the 4430.

BELOW: With increased hydraulic flow and power, it was helpful to be able to adjust the speed with which the hydraulics acted.

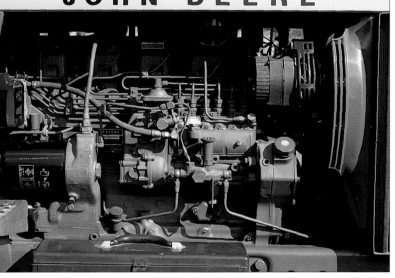

RIGHT: The Sound-Gard cab took on more of the look of an office. A comfortable seat, heating, air conditioning, and easy access to all controls took much of the drudgery out of farming.

Single rear wheels were inadequate to put the 4430's 126 horsepower to the ground. Dual rear wheels were common.

RIGHT: The 4430 had everything farmers were looking for, a combination of comfort, style, and a no-nonsense approach to a hard day's work.

ABOVE: Pumping 126 horsepower from 404 cubic inches of displacement, the turbocharged engine in the 4430 placed a lot of stress on the cooling and lubrication systems. Oil and water cooling systems had huge capacities.

Tricycle-type row-crop tractors, once the most popular style, were becoming things of the past when the 4430 came out. Row-crop tractors like this, with adjustable wide front axles and adjustable rear wheels, were the best sellers by 1976.

BELOW CENTER: The Sound-Gard body, introduced in 1972, provided unparalleled safety and comfort.

LEFT: Using both a turbocharger and an intercooler, John Deere squeezed 151 horsepower from its 404ci engine. It was becoming common for manufacturers to offer a single engine in different states of tune in different models.

BELOW: It took 37 years for John Deere to move from two-cylinder to six-cylinder engines, and it was another 22 years before they took the step to a V-8. This 8850 is powered by a 300hp, 955ci, turbocharged V-8 designed and built by John Deere.

By 1975 Deere felt it was time to fill out its small tractor line. The 830 and 1530 were replaced by the 40hp and 50hp 2040 and 2240, built in the Mannheim plant. Sixty and 70hp tractors, the 2440 and 2640, were built in the United States. The largest of the utility tractors was the German-built 81hp 2840.

In 1977 the mid-range got another overhaul. Five tractors were announced. The 4040 got its 90 horsepower from a 404ci normally aspirated diesel. The other three row-crop tractors shared the same 466ci six-cylinder engine in four states of tune. The 110hp 4240 used the engine in normally aspirated form. The 4440 and 4640 extracted 130 and 155 horsepower from the engine in turbocharged and turbocharged and intercooled form. The 6030's place was taken by the 4840, which for the first time offered the Sound-Gard cab on a 181hp tractor.

The escalation of horsepower left Deere with no tractor of less than 40 horsepower. There was a growing market for full-featured tractors of smaller size, so Deere turned to Yanmar of Japan to provide the 22, 27, and 33hp 850, 950, and 1050 tractors. These were powered by three-cylinder diesels, had 8-speed transmissions, and were available with mechanical front-wheel-assist.

The decade opened to a devastating recession in the agricultural economy. Many manufacturers couldn't weather the loss of business, and all suffered.

Deere's last new offerings of the decade were new versions of the 8430 and 8630 articulated tractors. The 8440 and 8640 had increased horsepower and greater comfort and reliability features.

The 1980s The decade opened to a devastating recession in the agricultural economy. Many manufacturers couldn't weather the loss of business, and all suffered. New models were introduced less frequently, as Deere sought to get the most out of existing inventory. Engineers were still at work behind the scenes, and in 1982 they had more efficient engine management systems, a revolutionary front-wheel-assist, and a new 15-speed Power Shift transmission ready. A new line of mid-size tractors was announced at this time. The 100hp 4050, 120hp 4250, 140hp 4450, the 165hp 4650, and the 193hp 4850. These were available with adjustable wide front axles or mechanical front-wheel-assist. The 4250 and the 4050 were available as a high-crop tractor.

Big tractors reached for even more power, and the 8440 and 8640 articulated four-wheel-drives were upgraded to the 187hp 8450 and 238hp 8650. The 8850, a 32,000-pound 304hp giant was added to the line, giving Deere three models of big articulated tractors.

The small tractor offerings were realigned, with all utility tractors coming from Japan or Germany. The very smallest tractors continued, but a 40hp model 1250 was added in 1982. In 1984, the 50hp 1450 and 60hp 1650 were added. The German-built tractors added in 1982 ranged, in 10 horsepower increments, from the three-cylinder 45hp 2150 to the six-cylinder 85hp 2950. These tractors differed from their predecessors primarily by having more horsepower and an 8-speed transmission with direction reverser. A special narrow tractor for vineyard and orchard work, the 50hp Model 2255, was also included.

Cane and rice growers as well as grove operators were relieved that 1985 saw an improvement in the farm economy and the introduction of several models with features these limited markets found indispensable. The

BELOW: Mechanical front-wheel-assist was a standard feature of the German-built 3155. This 95hp tractor was introduced in 1987. Standard equipment was a simple ROPS, but it could be fitted with a Sound-Gard cab.

RIGHT: The 8640 introduced the Investigator monitoring system that warned the operator of malfunctions in the engine and power train.

95hp 3150 was the first John Deere with mechanical front-wheel-assist as standard equipment. Also newly available were a low-profile 2750, and high-clearance 2750 with front-wheel-assist and wide-tread versions of the 2350, 2550, 2750, and 2950. A 25hp Model 900 tractor designed for cultivating a single row of flowers or vegetable was imported from Japan.

Another update of the line began in 1987 with the German-built models. The new models were the 45hp 2155, 55hp 2355, 65hp 2555, 75hp 2755, and 85hp 2955. These tractors were available with up to five choices of transmission and with mechanical front-wheel-assist. The 3155 with 95 horsepower replaced the 3150 in 1988.

Big tractors got a major upgrade the following year. The articulated tractors received an all-new chassis with a longer wheelbase and center frame oscillation. Three transmissions were offered, the 12-speed

One philosophy that has guided the company's tractor line...is the importance of economy. John Deere tractors have always been among the best at using fuel.

Synchro, 24-speed PowrSync and 12-speed Power Shift. Power for the 203hp 8560 and the 256hp 8760 came from John Deere six-cylinder engines. The biggest tractor with 332hp was the Cummins-powered 8960.

The last new tractors of the decade were a complete redesign of the row-crop line. All six new models came with a six-cylinder engine redesigned for improved fuel efficiency. The six models were the 4055 with 109 horsepower, 4255 with 123 horsepower, the 4455 with 142 horsepower, 4555 with 157 horsepower, the 4755 with 177 horsepower, and John Deere's first 200hp row-crop tractor, the 202hp 4955. All tractors were available with mechanical front-wheel-assist. The venerable 16-speed Quad Range transmission was standard, though Power Shift was an option. The 4255 was available on a high-crop chassis.

John Deere did not weather the 1980s unscathed, but its carefully-chosen tractor sizes and features along with wise management allowed it to end the decade with more than half the North American tractor market.

The 1990s The early 1990s presented more financial challenges to the tractor industry. Deere entered the decade cautiously, with few changes in the line. In 1992 the three larger row-crops were spruced up somewhat with a cleaner hood lines and renamed 4555, 4755, and 4955. The smallest tractor in this series, the 3155 was replaced by the 3255, a 100hp tractor with a new turbocharged engine. This basic tractor was also available without front-wheel-assist and the turbocharger as the 3055.

John Deere brought manufacture of some of its small tractor line back to the U.S. in 1992 when the 40hp 5200, the 50hp 5300, and the 60hp 5400 were announced. The German-built utility tractors were dropped, leaving no tractor in the 60 to 92 horsepower range.

ABOVE: The 4500, built by Yanmar of Japan and sold worldwide, is the largest tractor in John Deere's compact utility line.

This problem was corrected late in the year with the announcement of the 6000 series of mid-sized tractors. These tractors offered more big-tractor features than their predecessors. The Models 6200, 6300, and 6400 had 66, 75, and 85 horsepower respectively.

A new series of tractors replaced the 4055, 4255, and 4455 models. The new 7600, 7700, and 7800 were rated at 110, 125, and 145hp, an increase of five horsepower over their predecessors.

The big articulated tractors again got new engines and even more power. Cruise control allowed the tractors to maintain a constant ground speed in the field. An even bigger model was added—with an 855ci Cummins V-8, the 8970 produced 400 engine horsepower.

ABOVE: With front-wheel-assist and duals, this 145-horsepower 7800 is ready for any field conditions. It was also available in two-wheel-drive configuration.

In 1994 the last German-built tractors, the 3055 and 3255 were dropped, and the 7000 series was extended down to fill the power gap. The 7200, rated at 94hp and the 7400 at 102hp, both came standard with a 12-speed transmission. A 16-speed transmission, front-wheel-assist, and a choice of a roll-over protection frame or an enclosed cab were options.

By 1995 it was time to rework the larger tractors in the old 4000 series. They were given a new series name, 8000, and completely redesigned. A new 16-speed full powershift transmission was standard across the line. The 8100, 8200, and 8300 produced 160, 180, and 200hp. In keeping with John Deere's policy, these tractors shared the same basic engine in three states of tune. The largest in the line, the 225hp 8400, was equipped with a new 496ci engine. Front-wheel-assist was optional on the smaller tractors and standard on the 8400. Field Cruise control, first offered on the articulated tractors, was offered as an option.

More changes were in store before the end of the decade. The tractor line was organized into more distinct lines than previously. Japanese Yanmar imports comprised the 4000 series which ranged in power from 27 to 36hp. The 5000, 6000, 7000, and 8000 continued as described above. The articulated tractors were given their own series, the 9000s. John Deere responded to the success of Caterpillar's rubber-tracked farm tractors in 1997 by making the 8000 series available on rubber track chassis. In 1999 the tracked undercarriage was made available on the two larger tractors in the 9000 series, the 360hp 9300T and the 425hp 9400T. Deere got in one final update before the end

of the century, adding a "ten" to the series nomenclature. The new tractors all received increased engine efficiency and torque characteristics that minimized gear shifting. At the same time, Deere included its new fully automatic electronically controlled transmission as an option on the 6000 and 7000 tractors. A less sophisticated automatic transmission was available for the 8000.

At the turn of the century John Deere had weathered the challenges of 80 years in the tractor business, challenges that had spelled the end to once great tractor builders, and gone on to become the world's largest manufacturer of farm tractors. One philosophy that has guided the company's tractor line from the design of the first Model D in 1923 to the electronically controlled tractors of 2000 is the importance of economy. John Deere tractors have always been among the best at using fuel. In other ways John Deere has changed substantially. The company that once built the same model of tractor for 30 years has now stepped to the forefront of innovation. The 21st century will no doubt bring more innovation to the science of making farm tractors. Expect John Deere to be the leader in those innovations and, always, in economy.

ABOVE: *In 1995 the 200hp 8300 was available with two-wheel-drive or mechanical front-wheel-assist. In 1997 a track drive option was added.*

T H E   A M E R I C A N   T R A C T O R

★

MASSEY-HARRIS

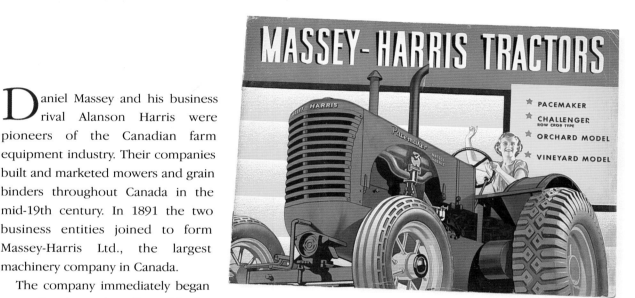

Daniel Massey and his business rival Alanson Harris were pioneers of the Canadian farm equipment industry. Their companies built and marketed mowers and grain binders throughout Canada in the mid-19th century. In 1891 the two business entities joined to form Massey-Harris Ltd., the largest machinery company in Canada.

The company immediately began expanding it product line with the intent of becoming a full line company. In 1911 the company introduced gasoline engines to its product range, the first step toward marketing a tractor.

By 1914 the company was ready, and began negotiations with the Bull Tractor Company of Minneapolis to market its tractor in Canada. The Bull was introduced to U.S. farmers in late 1913 at the sensational price of only $335. The Bull proved to be an unsatisfactory machine, and the company announced an improved Big Bull for 1915. It was the Big Bull that Massey-Harris marketed in Canada. The tractor had only three wheels. A wheel in front steered, a large wheel directly behind it was the drive wheel, and a small outrigger wheel to the left kept the tractor upright. It was rated at 24 belt and 12 drawbar horsepower, but had trouble living up to its rating. The tractor weighed only 4500 pounds, about half the average tractor of the time. The Big Bull was not a success in either the Canadian or U.S. markets, and by 1919 the Bull Company was no longer producing tractors.

ABOVE: Massey-Harris began selling the ill-fated Bull tractor in 1914. By 1918, it was building tractors on its own. The company introduced the first tractor designed and built in-house in 1930. From that time until the late 1950s, Massey-Harris remained a force to be reckoned with in the tractor business. Massey-Harris was also famous for building combines for the "Harvest Brigade" of WWII.

The First Massey-Harris The first tractor to carry the Massey-Harris name was designed in Illinois by Dent Parrett. Parrett had an intense interest in farm tractors and had designed several. By 1914 he had a model ready for production and had it manufactured for him by Independent Harvester of Plane, Illinois. In 1918 Massey-Harris arranged to build and sell the tractor in Canada under the Massey-Harris name. Like the Bull, the tractor was small and light in weight. It was a very simple design, just a steel framework with four wheels and a Buda engine mounted across the frame rails. Its transmission gave two forward speeds and one reverse. The Parrett/Massey-Harris was built in three sizes, the 12-25hp No.1, the 12-22hp No.2, and the 5800-pound 15-28hp No.3. The No.2 model was an improved No.1 and superseded it. Massey-Harris ceased production of the Parrett tractors in 1923, leaving it without a tractor to sell.

While Massey-Harris was working with Bull and Parrett, the Wallis Tractor Company of Racine, Wisconsin was developing a revolutionary tractor, the Wallis Cub. The Cub featured a unique unit frame constructed of a U-shaped length of steel plate. The U-frame enclosed the lower part of the engine and transmission, keeping them free of dust and dirt, and at the same time provided a light, rigid framework on which to build the

LEFT: *In the early part of the century, Massey-Harris had some of the most attractive and colorful advertising of its time.*

ABOVE: *This 1921 Massey-Harris Number 2, designed by Dent Parrett, was simply too crude to be successful against contemporary tractors like* the Fordson. *For all its faults, a Fordson was more durable, cost much less to buy, and was capable of any work the Massey-Harris could do.*

ABOVE: *As with its smaller stablemate the No.2, the design of the No.3 Massey-Harris was far behind that of the competition.*

tractor. This framework was the hallmark of Wallis, and later Massey-Harris, tractors for over 25 years.

In 1919 the Wallis Tractor Company and the J.I. Case Plow Works merged under the J.I. Case Plow Works name. The tractors retained the Wallis brand name. (The J.I. Case Plow Works is not to be confused with the J.I. Case Threshing Machine Company, a completely separate company with its own line of tractors covered in Chapter 2.) In 1926 Massey-Harris and the J.I. Case Plow Works entered an agreement where Massey-Harris would market Wallis tractors in Canada. The negotiations progressed beyond this initial agreement, and in 1928 it was announced that Massey-Harris would purchase the entire J.I. Case Plow Works.

Two tractors were initially offered, a 12-20 and a 20-30. Both were standard-tread tractors with four-cylinder engines and retained the unit frame first introduced in 1913. Massey-Harris continued to build them in the Racine, Wisconsin plant. For a short while the tractors were sold with the Wallis name on the radiator. Though by late 1929 they were identified as Massey-Harris products.

1930
Massey-Harris GP

By 1930 every serious tractor manufacturer was scrambling to develop a row-crop tractor. Massey-Harris's entry was a unique design called the General Purpose. The GP had four equal-size wheels which were all powered. The front wheels did the steering, and the rear half of the tractor rotated independently of the front to accommodate rough ground. Massey-Harris introduced a concept it called balanced traction, which placed most of the tractor's weight over the front wheels. A plow or other heavy draft implement attached to the rear tended to load the rear wheels, evening the weight distribution. Ironically, the General Purpose was a very good tractor for specific uses, but failed to live up to its name. It became a favorite for forestry, specialty crop farming, and other specific applications where a rugged wheeled tractor with excellent traction was needed. The GP balanced traction design was 30 years ahead of its time. The merit of the concept was not fully recognized until the 1950s, when four-wheel-drive tractors made their comeback.

MASSEY-HARRIS

LEFT: Early GPs were powered by Hercules engines, but in 1936 Massey-Harris began using its own power plant.

ABOVE: Its wide track and high clearance made the GP a popular substitute for a small crawler.

Most of the tractor's weight was cantilevered in front of the front wheels so that a draft load on the rear tended to equalize traction of the wheels. The chassis was quite flexible, with a pivot just in front of the rear axle allowing the front and rear of the tractor to rotate independently nearly 360 degrees. Wheel treads from 48 to 76 inches were available.

1930 MASSEY-HARRIS GP

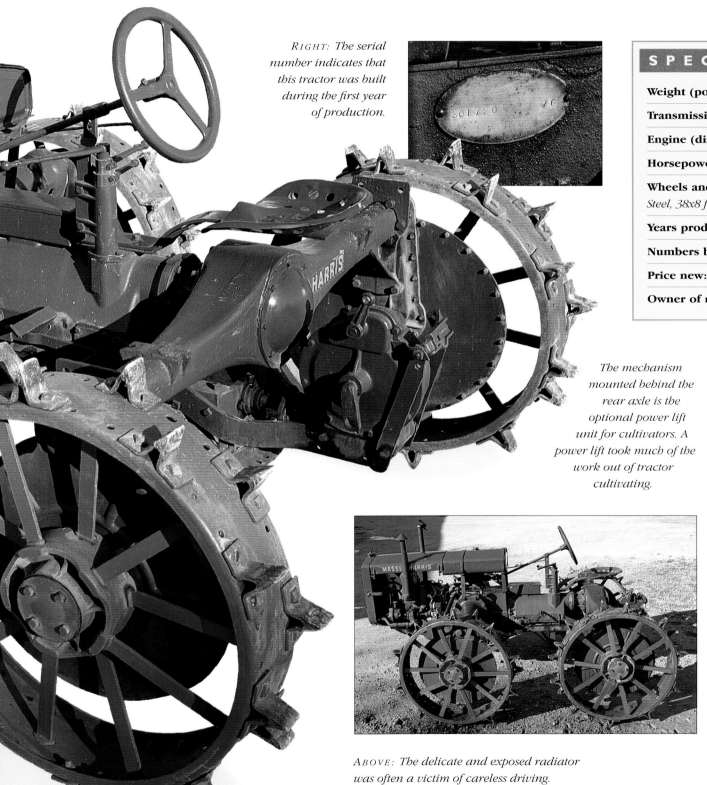

RIGHT: The serial number indicates that this tractor was built during the first year of production.

SPECIFICATIONS

Weight (pounds): *3861*

Transmission: *3-speed*

Engine (displacement, rpm): *226.2ci, 1200*

Horsepower (belt, drawbar): *22.20, 15.47*

Wheels and tires (front, rear):
Steel, 38x8 front and rear

Years produced: *1930–1938*

Numbers built: *c.3500*

Price new: *$1245*

Owner of machine pictured: *Bill Edgar*

The mechanism mounted behind the rear axle is the optional power lift unit for cultivators. A power lift took much of the work out of tractor cultivating.

BELOW: Massey-Harris designed and built its own oil bath air cleaner to help extend the life of the engine.

ABOVE: The delicate and exposed radiator was often a victim of careless driving.

TOP: *The Model 25 was the company's big standard-tread tractor throughout the 1930s.*

ABOVE: *Toys like this Challenger, built by Ertl, let Massey-Harris enthusiasts bring their hobby inside for the winter.*

BELOW LEFT: *The rear wheels of the Challenger were adjusted by sliding them on the axle, a method that was eventually adopted for nearly all row-crop tractors.*

purpose tractors in production and every serious tractor manufacturer was scrambling to develop one of their own. Massey-Harris's new tractor was an attempt at such a general purpose machine. In fact, they called it the General Purpose.

It was powered by a Hercules 226ci four-cylinder engine and had a 3-speed transmission. This tractor had four equal size wheels with power to all four wheels. The front wheels did the steering, and the rear axle could rotate independently of the chassis to accommodate extremely rough ground. The General Purpose had the required high ground clearance for row-crop work, but unlike other row-crop tractors, it had a wide turning

The 1930s In 1930 Massey-Harris announced a completely new tractor, the first to be a unquely Massey-Harris product. A huge market had opened for "general purpose," or "row-crop" tractors that could operate cultivators in row crops as well as pull heavy drawbar loads and operate belt-powered machinery. Farmall and John Deere both had general

"Clear Vision—Finger Tip Steering—Short Turning Make the Challenger a Great Tractor for Row-Crop Work"

– From a Massey-Harris Pacemaker and Challenger sales brochure

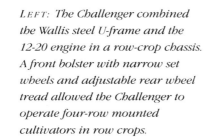

LEFT: The Challenger combined the Wallis steel U-frame and the 12-20 engine in a row-crop chassis. A front bolster with narrow set wheels and adjustable rear wheel tread allowed the Challenger to operate four-row mounted cultivators in row crops.

In 1932 the 20-30 was reworked to produce more horsepower and renamed the Model 25. It remained in the line through 1939. The Model 25 was primarily intended for heavy belt work, but was often pressed into service as a plowing tractor.

In 1936 a new attempt was made to produce a row-crop tractor using the 248ci engine and steel-plate unit frame of the 12-20. Called the Challenger, it followed a more conventional approach to the row-crop design. The Challenger had narrow front wheels and large rear wheels whose track could be adjusted by sliding them on long splined axles. The

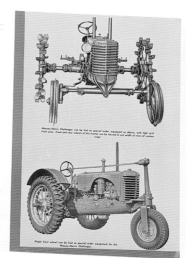

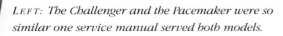

LEFT: The Challenger and the Pacemaker were so similar one service manual served both models.

ABOVE AND RIGHT: Orchard and vineyard versions of the Pacemaker were available. Demand for these specialty tractors was extremely limited. The few that exist are heavily sought after by collectors.

radius. This proved to be a serious liability. An additional shortcoming was the fact that the track was not adjustable, though six different wheel tracks from 48 to 76 inches wide were available from the factory. This limited the range of work an individual tractor was suited for. The General Purpose was a very good tractor for specific uses, but failed to live up to its name. This, combined with its relatively high selling price of over $1000, kept sales low. In 1936 it was given a new Massey-Harris-built engine and experiments were undertaken to fit a six-cylinder engine, but this was too little too late to make it a sales success.

*"Its high ratio of work output and low operating costs enable
users to meet readily the demands of war-time production."*

– From a Massey-Harris 101 Junior Tractor sales brochure

ABOVE: The 101 Junior was popular on small Canadian farms where, as in Europe, row-crop tractors had not gained great favor. The majority of 101 Juniors sold in the United States were row-crop types. The self-starter promised significant fuel savings.

RIGHT: Though the Model 22 had little in the way of advanced features, it could offer typical Massey-Harris reliability. It was a trim, handy little tractor, but its lack of a three-point hitch and live power take-off limited its appeal.

Challenger weighed 3700 pounds and produced 16-26 horsepower. The first Challengers were painted green with red wheels, but an improved model was introduced in 1937 that was identified by its red paint and yellow wheels. The improved tractor was called the Twin Power Challenger. It used the same engine, but the governor had two settings, one that could be engaged only when the transmission was in neutral that allowed an engine speed of 1400rpm, and the normal setting that allowed only 1200rpm. At the higher setting, the Challenger produced 41 horsepower.

A standard tractor, the Pacemaker shared the Challenger's engine and the Twin Power feature in a standard-tread chassis. Its development paralleled the evolution of the Challenger.

The Model 25 was painted red and given somewhat more stylish sheet metal for 1938. The engine was retuned to give more power, but the design was over ten years old and was obviously ripe for replacement.

New Models An entirely new line of tractors began to be introduced in 1938. The first was the Model 101. Though this tractor shared the Twin Power feature of the Challenger and Pacemaker, it was otherwise completely different. It was powered by a 201ci Chrysler six-cylinder engine that produced 31 or 36 horsepower at the belt and 24 at the drawbar. The famous U frame was gone. The engine was cradled in a heavy cast-iron framework with the 4-speed transmission case serving as the center of the frame and the rear axle housing forming the rear of the tractor's framework. Styling was a quite modernistic, with a grille that resembled the many-paned forward windows of contemporary aircraft and a fully enclosed engine. The 101 was available as a row-crop or a standard-tread tractor.

LEFT: The Model 81 standard saw work on farms and was also popular with the Canadian military as an aircraft tug.

In 1939 the 101 Junior was announced. It was similar in design to the 101, though at 3200 pounds it was 2000 pounds lighter and much smaller. The Junior was powered by a 124ci four-cylinder gasoline engine purchased from Continental. With 16 drawbar and 24 belt horsepower, it served a two-plow tractor market that Massey-Harris had ignored. It, too, was available as a standard-tread or row-crop chassis. Later, in 1946, a 102 Junior was offered. This model was the same as the 101 Junior except for its 162ci Continental engine equipped for distillate fuel.

An even smaller tractor was purchased from the Cleveland Tractor Company and sold through Massey-Harris dealers. The little tractor was called The General. A Hercules engine giving 11 drawbar and 17 belt horsepower drove the 2105-pound tractor. The General was a curious and short-lived product for the company. It was sold by Massey-Harris dealers, but it was never given the Massey-Harris name nor painted Massey-Harris colors. It was only sold by Massey-Harris for a couple of years, but went on to become a successful machine for other farm equipment retailers.

Like most tractor manufacturers, Massey-Harris ended the 1930s with new models in the line-up and excitement was high about what would come in the next decade. Unfortunately, the answer to that question was a long time coming, as the first half of the decade was consumed by war.

The 1940s Tractors reduced the amount of human labor required to operate a farm, and with so many men off fighting the war, they were in great demand. But building tractors took steel, rubber, and other materials that the war also demanded. Massey-Harris engineers scrambled to provide the most tractor power they could with the material available. The result was a flurry of similar models that remained in

ABOVE: The 101 Super was a sleek, modern tractor in its day. Its 218 cubic inch Chrysler engine produced 36 smooth horsepower.

RIGHT AND BELOW: The Model 44 was the most popular and successful of the postwar Massey-Harris tractors. It was available in standard-tread, row-crop, high-crop, rice-field, and orchard versions.

production only a short while. Because Massey-Harris was purchasing all its engines from outside sources, some models were built with different sizes and makes of engines.

The Model 25 was replaced with the 201 in 1940. With its cast-iron frame and stylish sheet metal, the 201 was similar in appearance to the 101. Its 242ci Chrysler engine produced 57 horsepower. The 201 came strictly as a standard-tread tractor, designed for the wheat- and rice-

ABOVE: The Model 55 was a brute of a tractor which acquitted itself well in western wheatfields.

growing territories. The 201 was adequate for pulling a four-bottom plow, but for the farmer who needed even more power, Massey-Harris offered the 202, essentially the same tractor with a 60hp Continental six-cylinder gasoline engine. A Model 203 was available with a 330ci, 64hp diesel engine.

In 1940 the 101 Junior was given a larger engine, a 140ci Continental, but was not renamed. For 1941 the 101 received an even larger 218ci Chrysler engine and became the 101 Super. A Model 81 was

"*An Easy Keeper...Massey-Harris Tractors deliver more horsepower hours per gallon of fuel...Make it a Massey-Harris*"

— Massey-Harris advertisement, 1950

announced in 1941. It used the same 124ci Continental engine as the first 101 Junior in a lighter, cheaper tractor. The 81 also found widespread use as an industrial tractor and airport tug. A companion tractor, the distillate-equipped Model 82, was also available.

Soon after the war Massey-Harris announced a revision of the whole line of tractors. The new tractors were similar to the successful prewar machines. Strong tractors with low power for their weight and little sophistication were the rule.

ABOVE: *The little Pony was painted gray and sold as a Ferguson as well as a Massey-Harris.*

The smallest of the new tractors was the Model 20. This little machine was an updated Model 81. A heavy two-plow Model 30 replaced the 101 Junior. The 30 was a 20-30hp tractor with a 5-speed transmission. Aimed at the popular two-plow class, it was only slightly more powerful than the Model 20, but at 3600 pounds it weighed nearly 900 pounds more, making it a much more robust two-plow tractor. Its 162ci four-cylinder engine was a standard Continental unit.

The next largest tractor, the Model 44, was one of Massey-Harris's biggest successes. A 260ci four-cylinder engine produced 31 drawbar and 40 belt horsepower for the 44. Engines tuned for gasoline or distillate fuel were initially offered, with LP and diesel engines soon available. It had a 5-speed sliding gear transmission and weighed 4100 pounds. It was available as a row-crop or standard-tread, with a high clearance version offered later. The 44-6 was also available with a six-cylinder, 226ci Continental engine of about the same power as the four.

Massey-Harris then turned to the smaller end of the line in 1947 and produced the Pony. This little one-plow tractor weighed only 1520 pounds and produced 8 drawbar and 10 belt horsepower. The Pony was used by tobacco and truck farmers or as a chore tractor on larger farms. It lacked sophisticated hydraulics, which limited its appeal. The tractor was built in Canada for the North American market and was powered by a Continental 62ci four-cylinder gasoline engine. A version was assembled in France that used European Simca or Hanomag diesel engines.

In 1948 the Model 22 replaced the Model 20. The 22 had a somewhat

ABOVE: *The Model 22 standard is not often seen in the United States, where the row-crop model was most popular. The standard was sold in substantial numbers in Canada and Europe.*

larger engine and produced slightly more power, but was otherwise similar to the 20. All the new tractors except the 55 were available with Depth-O-Matic, a rudimentary form of hydraulic hitch that did not include draft control.

As the 1940s closed, Massey-Harris could offer a tractor for almost any need. It had been among the first to offer diesel power in large and mid-sized tractors and was quickly developing a reputation for building rugged machines favored by farmers who used their tractors hard. Massey-Harris tractors did lack advanced features like draft-controlled three-point hitches, live power take-off, and live hydraulics, and a heavy three-plow tractor in the 35 horsepower range was conspicuously missing from the line.

BELOW: With both diesel and LP fuel options, the 33 lacked only live power take-off and a draft-sensing three-point hitch to put it at the top of three-plow class.

The 1950s In the showroom, at least, Massey-Harris appeared to be content with its tractor line-up. No new tractors were announced until 1952, when the small tractor line was reshuffled. The 22 was replaced by a new model called the Mustang. Another new model, the Colt, was little more than Model 22 with a smaller engine. The big news occurred in 1952 when the Model 33 replaced the 30. With a 201ci four-cylinder engine producing 28 drawbar and 34 belt horsepower, the 33 was the heavy three-plow machine that was so badly needed. Its 5-speed transmission and 4030-pound weight made it a three-plow tractor in the best Massey-Harris tradition. A diesel engine was offered for the first time. In 1954 the little Pony was given an increase in engine displacement to 91ci and renamed the Pacer.

The 44 was now the company's best-selling tractor. It came in row-crop, standard, and high-crop chassis and special equipment was available to adapt it to everything from rice field to orchard work. The 44 was then given slightly more power in 1953, creating some space between it and the new 33. The six-cylinder engine had been dropped in 1951, but the four-cylinder was available in gasoline, LP, distillate, and diesel forms, providing a engine for every need. With 43 belt and 34 drawbar horsepower, the updated 44 was called the 44 Special. More horsepower was also extracted from the gasoline-fueled Model 55, bringing the rated belt horsepower up to 58. Massey-Harris apparently did not think this change was special enough to warrant calling it the 55 Special.

ABOVE: Easily identified by the prominent hood ornament, the 333 added a 10-speed gearbox, more power, and improved hydraulics to the already successful Model 33.

In 1955 the Model 55 did get a name change, to the 555. Only detail changes were made to the old 55 to warrant the new name. The following year the 44 Special became the 444. A new dual range gearbox was added which gave the 444 ten forward speeds. The 33 became the 333 with the addition of the gearbox, but it also received a larger engine. The three-point hitches on both the 444 and 333 were improved. The triple digit tractors are easily identified by the large chrome ornament on the front of the hood.

The Massey-Harris organization had been experiencing difficulties, particularly with its small tractor line. The Pony, expected to sell in large numbers after World War II had proved to be a disappointment. The many small tractors introduced to patch up that situation were likewise disappointing because Massey-Harris lacked what the most successful small tractors had—a hydraulic three-point hitch with draft control. The

> ### *"Quick as a wink the Hydramic-Powered MH50 takes up the fight... drives through with extra torque, extra traction."*
> *– Advertisement for the Massey-Harris MH50, 1956*

company excelled at building large tractors and harvest equipment, but it took more than that to make a tractor manufacturer successful in the increasingly competitive market of the 1950s.

Another tractor company was struggling at this time as well. Harry Ferguson Inc. had a small tractor with all the right features, but the cost of manufacturing the machine made it unprofitable. Both companies were searching for a solution to their problems. On a July day in 1953 they found each other, and a possible solution.

Within weeks the two companies had agreed to merge and form Massey-Harris-Ferguson. Massey-Harris got a small tractor with a three-point hitch for its show-rooms. The tractor was the Ferguson TO-35, painted Massey-

Harris red and called the MH50. The 50 did differ from the Ferguson tractor in that it had a somewhat longer wheelbase and was available as a tricycle-type row-crop tractor. The rest of the Massey-Harris line remained the same and, with all the boardroom action taking place and demanding attention, the tractors were slipping behind technologically.

A management decision in 1956 called for selling off all the existing Massey-Harris tractors and putting the 50 into production. While production people built Massey-Harris tractors, sales people sold Massey-Harris tractors, and everyone looked forward to a brighter day, but the company's troubles continued. Managers approached John Deere with a mind to sell the company, but Deere declined to buy. In the end a decision was made to change the structure of the company entirely. A new corporation, called Massey-Ferguson, would come into being. It would build and market a completely new line of tractors under the Massey-Ferguson name (see Chapter 3). Those red and yellow tractors on the production lines and in the show rooms were the last Massey-Harris tractors there would ever be.

ABOVE: The gasoline engine and row-crop chassis was the most popular configuration of the 33. Thousands plied the fields of American farms.

TOP: The Mustang was an attempt at a small Massey-Harris. It was a step in the right direction, but the Ferguson merger doomed the entire small tractor line.

CHAPTER NINE

THE AMERICAN TRACTOR

★

MINNEAPOLIS-MOLINE

Minneapolis, Minnesota was a cradle of tractor development. More tractor manufacturers got their start in the Minneapolis area than anywhere else in the world. Minneapolis Steel and Machinery (MS&M) entered the tractor industry by building tractors on contract for other companies. The company introduced a tractor of its own under the brand name Twin City in 1910. Following conventional thinking, Twin City tractors were big. The first machine weighed over 25,000 pounds. Its engine displaced 1486 cubic inches and earned the tractor a power rating of 40 drawbar and 65 belt horsepower. A similar tractor with a 60-90 horsepower rating had a 2230ci six-cylinder engine and weighed 28,000 pounds. The company also made smaller tractors, of 25-45 and 15-30 horsepower.

An interesting addition to the MS&M line in 1917 was the Model 16-30. This little machine looked more like a large automobile than a farm tractor.

It stood no more than about four feet tall and the engine and operator station were enclosed in a sleek sheet metal body. Few were sold and it soon disappeared from the catalog. This would not be the last car-like tractor to be produced by this company.

In 1919 an advanced tractor with a unit frame and enclosed gearing was announced. This little 12-20 horsepower tractor would have been competition for the Fordson had the Fordson not sold for less than half its price. Its engine had four valves per cylinder and plenty of extra strength was built into every part. The machine was such an advanced design that it remained in production with minor changes for 15 years. Its only weakness was its transmission which had only two speeds. A larger tractor of similar design, the 20-35, was introduced in the same year. Both machines were re-rated in the mid-1920s, the smaller tractor became the 17-28 and the larger went to 27-44. The last new Twin City tractor was a 21-32, introduced in 1926. It was of similar design to the earlier tractors.

When the Minneapolis Threshing Machine Company, Moline Plow Company, and Minneapolis Steel and Machinery joined forces in 1929 to form the Minneapolis-Moline Power Implement Company, it was the MS&M machines that formed the nucleus of the new company's tractor line.

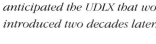

LEFT: The streamlined 16-30 of 1918 anticipated the UDLX that would be introduced two decades later.

ABOVE AND ABOVE RIGHT: Its location in the twin cities of Minneapolis and St. Paul, Minnesota prompted the Minneapolis Steel and Machinery company to adopt the name Twin City for its line of tractors.

"Dependable as the locomotive—
Twin City 12-20 Kerosene Tractor"
– Minneapolis Steel & Machine Company advertisement, c.1920

RIGHT: MS&M jumped from building gigantic gas traction engines straight to modern, lightweight tractors in 1919. The 21-32, with minor updates, remained in production for 15 years.

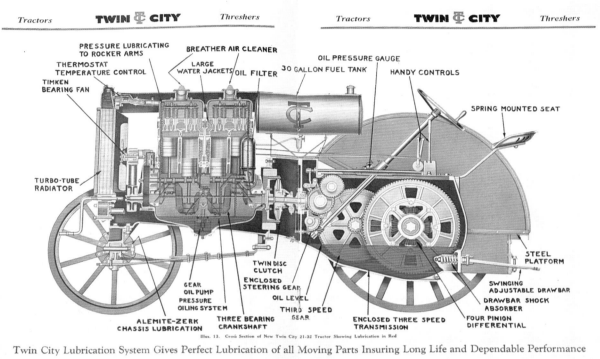

| Tractors | TWIN Ⓣ CITY | Threshers | | Tractors | TWIN Ⓣ CITY | Threshers |

PRESSURE LUBRICATING TO ROCKER ARMS
BREATHER AIR CLEANER
THERMOSTAT TEMPERATURE CONTROL
LARGE WATER JACKETS
OIL FILTER
OIL PRESSURE GAUGE
30 GALLON FUEL TANK
HANDY CONTROLS
TIMKEN BEARING FAN
SPRING MOUNTED SEAT
TURBO-TUBE RADIATOR
STEEL PLATFORM
TWIN DISC CLUTCH
GEAR OIL PUMP PRESSURE OILING SYSTEM
ENCLOSED STEERING GEAR
OIL LEVEL
THIRD SPEED GEAR
SWINGING ADJUSTABLE DRAWBAR
DRAWBAR SHOCK ABSORBER
ALEMITE-ZERK CHASSIS LUBRICATION
THREE BEARING CRANKSHAFT
ENCLOSED THREE SPEED TRANSMISSION
FOUR PINION DIFFERENTIAL

Illus. 13. Cross Section of New Twin City 21-32 Tractor Showing Lubrication in Red

Twin City Lubrication System Gives Perfect Lubrication of all Moving Parts Insuring Long Life and Dependable Performance

MINNEAPOLIS STEEL AND MACHINERY COMPANY
Page Eight

MINNEAPOLIS STEEL AND MACHINERY COMPANY
Page Nine

Another
TWIN Ⓣ CITY
TRACTOR

The NEW 21-32

MINNEAPOLIS STEEL and MACHINERY CO.

LEFT AND BELOW: Twin City tractors ruled the prairies in their day. Their size and smooth power made them favorites with those large farms and big custom threshing operations that could afford them.

LEFT AND BELOW LEFT: Though Minneapolis-Moline retained the Twin City name on tractors that were in production when the company was formed in 1929, new model numbers were assigned when updates were made. The Model 21-32 became the FT in 1932.

BELOW: The Model MT was Minneapolis-Moline's first tricycle-type row-crop tractor. It proved to be too big for the largest segment of the row-crop market.

Soon after Minneapolis-Moline was formed, a row-crop tractor was announced. This machine was obviously under development at MS&M before the merger. Like the John Deere GP, the Model KT followed the three-row cultivating theory, with the tractor straddling only one row. The KT was a robust little tractor, weighing 4300 pounds for its 15/23 horsepower rating.

The 1930s At the start of the Depression decade, Minneapolis-Moline had five tractors in its line. the Minneapolis 27-42 as well as the Twin City KT, 27-44, 21-32, and 17-28 tractors The Minneapolis 27-42 did not last long, as the Twin City 27-44 was a superior tractor of similar size. A year later the MT, a tricycle-type tractor using the engine and transmission of

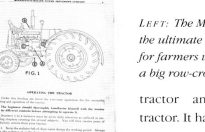

the KT, was announced. With the MT, Minneapolis-Moline had the row-crop market covered no matter whether farmers ultimately made three- or four-row cultivation the standard. 1932 brought the Model FT, an improved 21-32.

As the decade progressed, interest in the larger tractors waned. Threshermen who used big tractors like the 27-44 liked them just as they were, so improvements came slowly. Nineteen thirty-four, however, saw a sweeping change in the smaller tractors. A brand new tractor was announced. This little machine, the Model JT, was a tricycle row-crop intended to compete with the likes of the Farmall F-20 and John Deere A. It was also available as an orchard

LEFT: The Model U was the ultimate in power for farmers who needed a big row-crop tractor.

tractor and a standard-tread tractor. It had a 196ci four-cylinder engine producing 14/22 horsepower and a 5-speed transmission. Other tractors in the line were upgraded at this time also. The KT and MT were retuned and produced 19/30 horsepower as the KTA and MTA. The FT received an increase in engine displacement from 382 to 403ci resulting in a power increase to 27/41. It was renamed the FTA.

In 1936 the first of a striking new line of tractors for Minneapolis-Moline was announced. The Universal Z (or ZTU) was a row-crop tractor with an all-new engine of 185ci, and a 5-speed transmission. The engine had a unique cylinder head design in which the valves opened by long

ABOVE: The little Model JT was sized right for the average row-crop farmer, but technology was moving forward rapidly. Almost as soon as it appeared, Minneapolis-Moline was at work on a more modern tractor, the Model R.

rocker arms which moved them horizontally. This simplified maintenance for do-it-yourself farmers. Stylish sheet metal covered the engine and radiator and tapered sharply back to a narrow fuel tank. The narrow tank allowed excellent visibility for the operator of front-mounted cultivators. The old gray color scheme of previous tractors was gone, replaced by rich Prairie Gold yellow with red wheels. A standard-

"Again Minneapolis-Moline leads the parade!
With the new Standard 'U' tractor."

– 1938 Minneapolis-Moline advertisement for the Standard U

tread version, the ZTS, followed in 1937. The Z replaced the Model J tractor, which was slightly under-powered.

Minneapolis-Moline had a unique way of naming its new line of tractors. Each tractor name consisted of three letters. The first was the series, the second was always a T and stood for "tractor." The final letter indicated the chassis type. Universal was the nomenclature Minneapolis-Moline adopted for row-crop tractors. So the ZTU was a Z series row-crop tractor. The ZTS was a Z series standard-tread tractor.

In 1938 the Model U series was announced. This tractor was styled like the Z, but at 38 horsepower, was 65 per cent more powerful. In row-crop form, the UTU was an impressive machine. It weighing 5250 pounds and was capable of pulling a four-bottom plow in most conditions. Like the Z, it was available in both standard and row-crop configurations. A variant of the U, the UDLX, was a sensationally futuristic tractor with full fenders, a fully enclosed steel cab, heater, radio, dual

ABOVE: The Model R proved to be a popular row-crop tractor.

seats, and a road speed of 40mph. The UDLX could be used to plow in comfort all day and driven to town on Saturday night. Its $2150 price tag and farmers' customary frugality kept it from being popular. Only about 150 were sold.

The Model GT was also announced in 1938. The GT used the same engine that had been designed for the 21-32 in 1926 in 403ci form. At 6220 pounds the GT was a massive tractor that produced 36 drawbar and 49 belt horsepower.

The last new tractor in the new line was the R series of 1939. The R was a smaller version of the Z, producing 18/24 horsepower. It used a 165ci engine similar to the engine in the Z and had a 4-speed transmission. Like the Z, it was available as a row-crop, wide front row-crop, and a standard-tread tractor. An industrial version was also made and a steel cab was offered.

The 1930s ended with four modern tractor series, the G, U, Z, and R in the line. These four sizes provided platforms on which the company could build a strong future.

LEFT: Minnie Mo was M-M's spokesperson in the late 1930s. She exemplified the ease with which Minneapolis-Moline's power could be controlled.

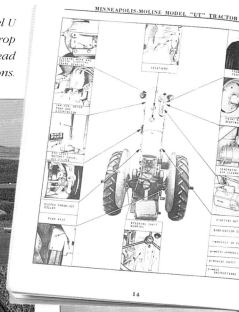

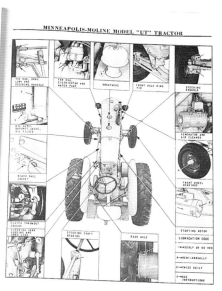

RIGHT: The Model U came in both row-crop (UTU) and standard-tread (UTS) configurations.

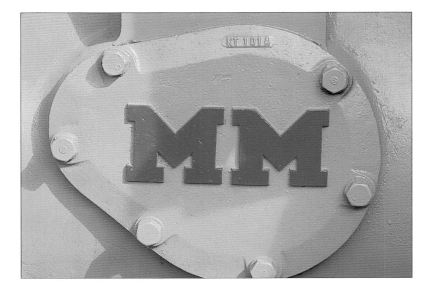

ABOVE: With 43 maximum horsepower and weighing 5500 pounds, the UTU was the largest row-crop tractor available in 1938.

RIGHT: The RTE was the standard-tread version of the Model R. It came with full fenders and a flat operator's platform.

ABOVE: To emphasize the strength and power of Minneapolis-Moline tractors, the company chose heavy block letters for its logo. The M-M logo was cast into many of the tractors' parts.

1938
Minneapolis-Moline UDLX Comfortractor

Forming the sheet metal that covered farm tractors into streamlined shapes had become standard practice by the late 1930s. But the Minneapolis-Moline UDLX took the concept several steps further—it was more than a pretty face. When it was first shown to the public, the "Comfortractor" caused a sensation. With its fully enclosed steel cab, heater, radio, dual seats, and a road speed of 40mph, the UDLX could be used to plow in comfort all day and driven to town on Saturday night. While it was not a commercial success, only about 150 were ever sold, it served as a harbinger of what was to come. Using tractors for transportation never became as popular in America as it was in Europe, but farmers liked the idea of having comfort and protection from the elements. Within a few years factory-built cabs became available from manufacturers, and by the late 1990s a comfortable, heated, air-conditioned cab with a stereo sound system and a cell phone was de rigueur.

RIGHT: The operator's station was a more like that of a truck than a farm tractor.

LEFT: A brake and tail light was provided for operation on the road.

Not all UDLX tractors came with a cab. A sportier version without the top was available, though very few were made. Would this be considered a roadster or a convertible?

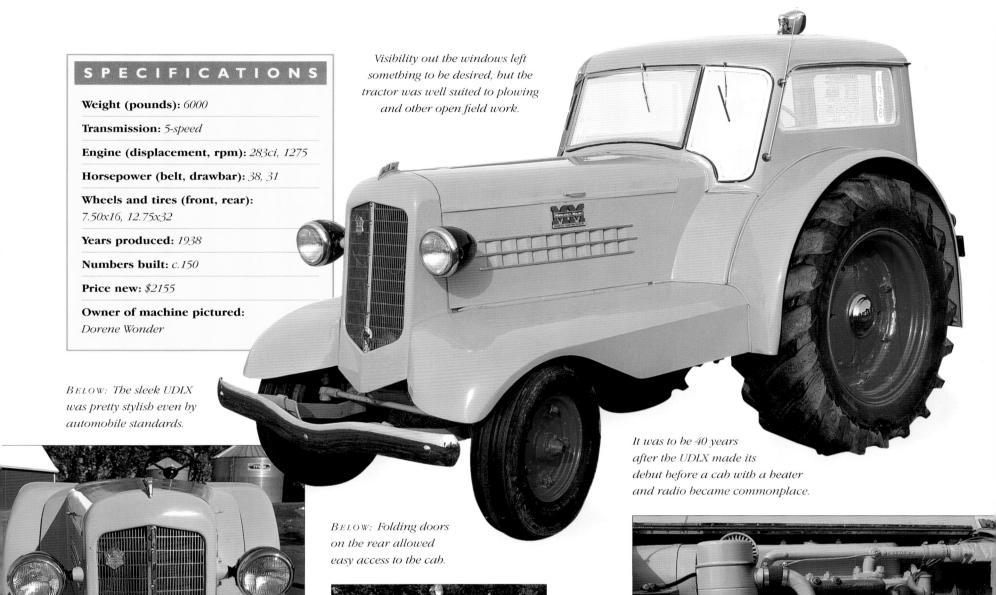

SPECIFICATIONS

Weight (pounds): *6000*

Transmission: *5-speed*

Engine (displacement, rpm): *283ci, 1275*

Horsepower (belt, drawbar): *38, 31*

Wheels and tires (front, rear):
7.50x16, 12.75x32

Years produced: *1938*

Numbers built: *c.150*

Price new: *$2155*

Owner of machine pictured:
Dorene Wonder

Visibility out the windows left something to be desired, but the tractor was well suited to plowing and other open field work.

BELOW: The sleek UDLX was pretty stylish even by automobile standards.

It was to be 40 years after the UDLX made its debut before a cab with a heater and radio became commonplace.

BELOW: Folding doors on the rear allowed easy access to the cab.

ABOVE: Hidden under its sleek hood, the UDLX employed the same 283ci four-cylinder engine that powered the Model U.

"Just what we need for this land of ours."
– *1941 advertisment for the Minneapolis-Moline model range*

ABOVE: The GTA was the largest tractor in the Minneapolis-Moline line.

BELOW: This 1947 GTA is equipped to burn liquid petroleum (LP) gas. Minneapolis-Moline was a pioneer in engineering engines for LP fuel.

The 1940s Minneapolis-Moline was heavily involved in building vehicles for the war effort and made few changes to the tractor line. After peace came, the company concentrated on improving its existing tractors rather than bringing out new models. In 1942 the GT was renamed the GTA. The grille was painted yellow instead of red, but the tractor was unchanged otherwise. In 1947 the GTA was substantially improved and became the GTB. With a 403ci engine and 5-speed transmission, the GTB cranked out 39 drawbar horsepower and 51 horsepower at the belt.

A lot of work was done on engines and fuels. The Model U was offered with factory-installed liquid petroleum gas (LP) fuel equipment in 1941. A Minneapolis-Moline was the first LP-powered tractor to be evaluated at the University of Nebraska tractor-testing laboratory. The tractor was a Model U equipped with a high-compression engine of the same displacement as the gasoline and distillate versions. LP fuel was inexpensive in the oil-producing territories, and its well-

ABOVE: The little Model V, adopted when Minneapolis-Moline bought Avery, was the company's smallest domestic tractor.

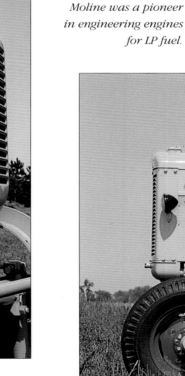

LEFT: *Minneapolis-Moline built many tractors for military use.*

BELOW: *The little Model V remained in the Minneapolis-Moline line for only two years.*

engineered liquid petroleum gas engines made Minneapolis-Moline tractors favorites in these areas. An improved Model Z, the ZA, was introduced in 1949. With a displacement of 206ci, the new ZA produced 25 drawbar and 32 belt horsepower.

Minneapolis-Moline Power Implement Company officially changed its name to Minneapolis-Moline Company in 1949. The Twin City brand name of its tractors was also officially dropped.

The 1950s The new decade saw much more activity from Minneapolis-Moline; unfortunately for the rest of the tractor line this activity was not focused on the company's mainstream tractors, but rather on the Uni-Tractor, a simple chassis powered by a tractor engine that looked somewhat like a row-crop tractor with the driver facing backwards. The harvesting equipment—hay balers, combines, forage harvesters, and corn pickers etc—that were especially

LEFT: *Minneapolis-Moline built many tractors for military use.*

1943 ZTX · MIL 224

ABOVE: *Minneapolis-Moline advertising stressed the advantages of the streamlined shape of its tractors.*

Diesel power was the coming thing, and Minneapolis-Moline was not going to be caught unprepared.

LEFT: Calendars reminded farmers of Minneapolis-Moline daily.

designed for it were mounted directly on the unit. The company had no small tractor, and had sat by and watched as the small tractor market grew in the 1940s.

In 1950 Minneapolis-Moline acquired the B.F. Avery Company, a manufacturer of light tractors and equipment in Louisville, Kentucky. B.F. Avery had a network of dealers across the southeast, where Minneapolis-Moline was barely known, as well as two small tractors that had been popular with the small acreage farmers. Minneapolis-Moline added the tractors, the one-plow Model V and the light two-plow Model BF to its line. The V, weighing 1612 pounds and powered by a 65ci Hercules engine, was a popular cultivating tractor among vegetable and tobacco farmers. The 2880-pound, 132ci BG was available with optional axle and wheel combinations to adapt it to row crops, high-growing crops, and vineyard use.

With a light two-plow tractor in the line there was room to increase the power of the Model R to make it a heavy two-plow machine. In 1951 a high compression gasoline model was tested at Nebraska that produced 18/24 horsepower. An LP fuel engine was developed for the big GTB in 1951. It was called the GTC when it was equipped with this engine.

Diesel power was the coming thing, and Minneapolis-Moline was not going to be caught unprepared. A few Model Us were built in 1952 with engines converted to diesel power. The next year the UB was introduced. This tractor had new, more convenient operator's station, new grille and sheet metal, and a foot clutch. The ZB was also announced in 1953. It had the same running gear as the previous model but, like the UB, it got an improved operator's station and a new, squared-off grille.

BELOW: M-M developed a dual line of mounted harvesting equipment, one line for tractors and one for the Uni-Tractor.

BELOW: The RTE, with adjustable wide front axle, is one of the rarer variants of the R series.

LEFT: The Uni-Tractor was next to useless without its special mounted equipment. Its principal value was to cut the cost of self-propelled equipment and to free a tractor to do other work at harvest time.

A big 425ci diesel engine became available for the Model GTB chassis in 1953. When equipped with this engine the tractor was called the GTB-D. It was produced for only two years before being updated and renamed the GBD. The GB continued through 1959.

Minneapolis-Moline seemed ready to get out of the small tractor field almost as quickly as it entered it. The Model V was dropped in 1953 in favor of the BG, a BF modified for one-row cultivation work. The entire BF series was subsequently dropped. The old Model R was dropped the next year, leaving Minneapolis-Moline without a tractor of less than 30 horsepower.

The GTB was updated in 1955. When equipped with a gasoline or LP engine, the same engine as used in the GTB and GTC, this tractor was known as the GB and produced 46/58 and 50/64 horsepower respectively. The diesel used the 426ci engine and produced 44/56 horsepower. The GB was the first Minneapolis-Moline tractor to offer live power take-off.

New Tractors Minneapolis-Moline introduced two completely new tractors for 1955. After half a decade of slipping slowly behind, Minneapolis-Moline now had two tractors that were as modern and up-to-date as any on the market. The 335 was a utility-type tractor with adjustable front axle and rear wheel tread. Its 165ci four-cylinder engine was only available as a gasoline model and produced 24 drawbar and 30 PTO horsepower. The transmission was a 5-speed sliding gear type with Ampli-Torc, which gave a total of 10 forward speeds. Ampli-Torc was a 2-speed planetary gearset that allowed shifting between two adjacent speeds without stopping or declutching. A three-point hitch with draft control and live PTO were both offered. The 445 was a larger tractor, available as a row-crop or utility type. It was powered by a 206ci four-cylinder engine and produced 31 drawbar and 38 PTO horsepower. Like the 335, Ampli-Torc, live hydraulics with three-point hitch, and live PTO were available. The 445 could was available equipped for gasoline, LP, or diesel fuel.

The venerable Model U line was replaced by a rugged new four-plow 5-Star in 1957. The gasoline and LP versions of the 5-Star was powered by the 283ci UB engine. The diesel engine was a version of the 283 enlarged to 336ci displacement. The 5-Star was available as a row-crop or standard-tread tractor. All of the new tractors got attractive styling and a revised operator station that placed the driver ahead of the rear axle and astride the transmission. The line was completely revamped in 1959, with new styling and new color schemes.

Minneapolis-Moline demonstrated its lack of understanding of corporate identity by making sweeping changes in the model line, but continuing to brand its tractors in an inconsistent manner. The smallest tractor was the Jet Star, which replaced the 335, but had a larger engine and more power. It was painted brown. The 445 was replaced by the 4-Star and was painted Prairie Gold and brown. The replacement for the 5-Star was called the M5 and was painted to match the 4-Star. The largest tractor in the line was called the Gvi and was painted the same as the M5 and 4-Star. With the change, the smaller tractors in the line all received small power increases. The new Gvi offered a higher operator's station, more power, and modern styling on what was the GB chassis.

ABOVE: The 335 was Minneapolis-Moline's first entry into the utility tractor category. M-M wisely adopted Ferguson's three-point hitch right off instead of trying to interest farmers in its own version.

Minneapolis-Moline ended the 1950s with its tractor line tending in a positive direction. Hydraulics, horsepower, and operator comfort were all keeping up with the competition. But there was still no tractor of under 30 horsepower, which put dealers at a disadvantage and weakened the company.

RIGHT: The handy little U302 was introduced in 1964.

BELOW: M-M engines had plenty of reserve strength, and could stand up to being hot-rodded. This LP model with a turbocharger added could put out over 110 horsepower.

The 1960s The 1960s began with Minneapolis-Moline putting ever more power to the soil. Small tractors were left to the other companies, as bigger engines and four-wheel-drives were developed. In 1962 Minneapolis-Moline announced the industry's first modern small (under 100 horsepower) four-wheel-drive tractor. The M504 was an M5 with a mechanically driven front axle. The next year saw a mechanical front-wheel-drive for the Gvi, called the G704, being offered. Also in 1962, the Jet Star 2, a bargain-priced Jet Star with few options, was added to the line.

The following year the Jet Star, Jet Star 2, and 4-Star were all dropped and replaced with the Jet Star 3. The Jet Star 3 was a 4-Star with new squared-off sheet metal. A Super version with more amenities for the operator was available.

Two new big tractors were announced in 1963, the G705 and G706. The G705 replaced the Gvi. It was a similar tractor with a 504ci six-cylinder engine that produced 101 horsepower in LP form or 102 in diesel form. A gasoline version was not available. The G706 added mechanical four-wheel-drive to the G705 chassis. These were the first to carry Minneapolis-Moline's newly styled sheet metal with more attractive, rounded corners. In 1963 the M602 and M604 replaced the M5 and M504. Except for the new sheet metal styling, these tractors were little different from the models they replaced.

In 1963 Minneapolis-Moline was purchased by White Motors, but the company continued to bring out new models under the Minneapolis-Moline name. From 1963 on, Minneapolis-Moline was controlled by the managers of White Motors. Decisions were based on what was best for White Motors, resulting in Minneapolis-Moline's steady decline.

The new "little" tractor in the Minneapolis-Moline line was announced in 1964. This, the U302, was an updated Jet Star 3 with a larger engine, 56 horsepower, improved hydraulics, and new sheet metal that matched the larger models. A row-crop model as well as a utility were offered.

Minneapolis-Moline offered a new row-crop in 1965. The G1000 put the big 504ci diesel and LP engine in a row-crop chassis. The same 5-speed with Ampli-Torc transmission was offered. In 1967 the operator's station was moved up and forward and the fuel tank was moved to behind the seat. The tractor was renamed the G1000 Vista. A smaller version of the G1000, the G900 with a 425ci engine equipped for gasoline, diesel, LP also became available in 1967. All were available as wheatland tractors with a solid front axle.

Minneapolis-Moline continued building new tractors from the old parts bins in 1969. A tractor built like the old G1000 was given new sheet metal and called the G1050. The G900 received similar treatment to become the G950.

A big articulated four-wheel-drive tractor, the A4T 1400, was announced by Minneapolis-Moline in 1969. The tractor used the same six-cylinder diesel engine as the G1000 Vista and a 10-speed transmission.

Both in terms of prestige and product, Minneapolis-Moline had declined precipitously in the 1960s. It ended the decade with five tractors in the line. In 1963 one of Minneapolis-Moline's greatest achievements, the Uni-Tractor, had been sold off to New Idea. Treated as a merely a unit of White Motors, Minneapolis-Moline was now obviously being unceremoniously relegated to the sidelines.

The 1970s The 1970s saw White badge-engineering its product lines heavily. With a different grille and green and white paint the Minneapolis-Moline G1050 was sold as the Oliver 2055. The G950 was also the Oliver 1865. White added smaller tractors to the Minneapolis-Moline line, but under the paint these were identical to Oliver tractors. The 41 and 59hp

In 1974 White ended the agony of Minneapolis-Moline loyalists by dropping the Minneapolis-Moline name entirely.

ABOVE: The G1050 used the trusty M-M six-cylinder engine in its 504ci size. It was also sold as the Oliver 2055 in 1971.

utility tractors, called the G350 and G450 when in Minneapolis-Moline trim, were Fiats, imported from Italy, repainted, and rebadged for the Minneapolis-Moline, Oliver, or Cockshutt dealer White was sending them to.

There was still some life in the Minneapolis-Moline line, however. The venerable GB six-cylinder engine was enlarged again, to 585ci, and placed in the G1050 chassis to create the G1350. Planetary final drives were added to relieve some of the stress from the old gear train, as the big diesel put out 135 horsepower. This tractor was offered only in 1972. Minneapolis-Moline had fallen behind in power train development, but White stablemate Oliver had a newer 18-speed transmission and final drive, so the 585ci Minneapolis-Moline engine was wedded to Oliver's transmission to create the 143hp Minneapolis-Moline G1355. This tractor was also available with a 504ci engine equipped for LP fuel. In 1973 a 451ci version of the GB engine was matched to an Oliver power train that offered a 6- or 18-speed transmission and became the G955

While other manufacturers were abandoning every fuel type but the diesel, Minneapolis-Moline was faithful to the farmer who preferred to power his tractors with clean LP fuel. In 1970 Minneapolis-Moline offered the only LP-powered articulated tractor in the industry, the 169hp, 17,300-pound A4T 1600 LP. A diesel version was available that produced 143 horsepower.

The Minneapolis-Moline content in tractors that carried the Minneapolis-Moline name had steadily diminished from the late 1960s through the early

1970s. In 1974 White ended the agony of Minneapolis-Moline loyalists by dropping the Minneapolis-Moline name entirely. Though not acknowledged by White, Minneapolis-Moline's legacy lived on in the engines it designed in the 1950s, which continued to be used in White brand tractors for several years. Not content to let one of the grand names in American agriculture pass with dignity, in 1989 White heaped a final insult on Minneapolis-Moline when it announced the American 60 and 80 series tractors. In an attempt to capitalize on the brand loyalty that to this day remains strong among owners of old Minneapolis-Moline tractors, it applied Minneapolis-Moline paint colors to generic tractors that had no Minneapolis-Moline content at all. Demonstrating its lack of understanding of the American farmer, White insulted Cockshutt and Oliver enthusiasts by painting the same tractors Oliver green and Cockshutt red. Today, Minneapolis-Moline tractors continue to work in American fields, and through the efforts of thousands of loyal Minneapolis-Moline collectors, they undoubtedly will continue to do so for many decades.

RIGHT: The A4T, which was available with 139 to 169 horsepower, was also sold as the Oliver 2455 in the U.S. and as the White Plainsman in Canada.

CHAPTER TEN

THE AMERICAN TRACTOR

ORPHANS AND OTHERS

Over the years there have been many tractor manufacturers who have produced a successful and influential product, but failed to sustain their success over a period of time. Often the failure was due to mismanagement. Sometimes the company was too small to resist a buyout from a larger company that wanted to add their innovation to its line. In some instances the company only had one good idea in it, and failed when advances in technology passed this innovation by. The saddest cases are those where corporate greed and connivance worked to wrest the company away from hard-working and dedicated owners.

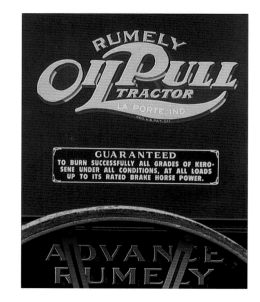

ABOVE: The Rumely OilPull excelled at burning cheap fuel.

Advance Rumely

The Advance Rumely company built dozens of products, from threshing machines to motor plows, but is best known for the OilPull tractor. The first production OilPull hit the fields in 1910. The tractor was huge, weighing around 25,000 pounds, and produced about 25 drawbar and 45 belt horsepower. It was unique in its ability to burn cheap kerosene fuel effectively under any conditions. Kerosene and other heavy fuels burn efficiently within a narrow range of temperatures. If the cylinder is too hot, the fuel ignites prematurely. If it is too cool, the fuel does not burn completely. The OilPull used oil as a cooling medium. This allowed cylinder temperatures to remain high enough to burn kerosene without boiling away the coolant. It was also fitted with a patented carburetor that mixed water with the fuel as needed to cool the cylinder when temperatures were too high.

With one exception, all OilPulls shared the same basic design. A large, slow-turning two-cylinder engine and a transmission were mounted on a frame. Open gears drove large steel wheels, and a cooling tower was mounted at the front. The tractors were adequate plow tractors, but they really excelled at belt work. Their heavy, smooth-running engines provided steady power to a thresher or sawmill.

Early OilPulls were known as the heavyweights, and came in six sizes; the B, 25/45hp, built in 1910 through 1912; the E, 30/60hp, 1910–1923; the F, 18/35hp, 1911–1918, the G, 20/40hp, 1918–1924; the H, 16/30hp, 1917–1924; and the little Model K, 12/20hp built from 1918–1924. The only exception to the basic OilPull design was the Model F, which used a one-cylinder engine built with components from the Model E. Though it was it was not the biggest seller, the Model E was a grand, imposing machine. In 1911 Rumely demonstrated the power of its tractor by chaining three Model Es to a 50-bottom plow. The rig plowed an acre of ground every four and a half minutes and in six days plowed over 2000 acres.

In 1924 the OilPull line got its first overhaul. The old steel I-beam frame was replaced with a lighter pressed steel one, engines were updated, and a new, entirely enclosed transmission was used. The spoked flywheel was abandoned in favor of a solid plate

ABOVE: Most farmers found that the 16-30 was the right size for their operations. Rumely built over 10,000 of them.

LEFT: The 30-60 E was the first commercially successful OilPull. Though it sold relatively few each year, the 30-60 was very well regarded by farmers who needed a big tractor, and demand for it was steady. It remained in production for 13 years.

BELOW: When farmers called for a good, simple, cheap tractor, P.J. Lyons offered them two out of three of those requirements with the Little Bull. Farmers quickly learned that, while it was simple and cheap, unfortunately the Little Bull wasn't much good.

type. Engine speeds were turned up, so each size produced more power, and the 2-speed transmission was replaced with a 3-speed. The lightweights remained in the line with some changes through 1928 when engine modifications added about 20 per cent more horsepower. The new tractors, called the Super Power line, were the ultimate in gas traction engines, but throughout the 1920s the big gas traction engines lost favor to lighter more nimble tractors. Rumely developed a general purpose tractor and a modern unit frame tractor by 1929, but they were too late to save the company. Allis Chalmers bought Rumely in May 1931 and liquidated all Rumely products.

The Bull Tractor Company

The Bull Tractor Company is noted for its negative impact on the tractor industry. No company enjoyed such a rapid rise in its fortunes, and few fell so abruptly. Bull was established in January 1914 by P.J. Lyons. By the end of that year it had sold 3800 tractors, enough to place it at the top of the tractor industry in sales. The company had one product, a tractor called the Little Bull. It was an absurdly simple three-wheeled machine with only one drive wheel. Power was provided by a two-cylinder 230ci engine. The

transmission consisted of simple gearing that only provided one forward and one reverse speed. The tractor was rated at five drawbar horsepower and 12 belt horsepower.

The Little Bull promised exactly what thousands of American farmers clamored for—an inexpensive tractor small enough for the average American farm. Unfortunately, it failed to deliver on its promise. By the time the tractor had been on the market for a full season, farmers learned

1918
Rumely OilPull 16-30

OilPull tractors were one of the most thoroughly developed of the early, heavyweight farm tractors. The engine operated on the principle of maintaining extremely high cylinder temperatures, which allowed it to ignite almost anything. Though it consumed it in prodigious quantities, the OilPull would burn the cheapest of liquid fuels. To maintain the high temperatures it was oil-cooled, which also eliminated rust and freezing problems common in tractors of its day. A special and highly successful carburetor was developed that automatically mixed water with the incoming fuel to keep cylinder temperatures from getting too high. OilPulls had developed such a following among threshermen and sawmill operators that dealers were still selling 15,000-pound OilPulls ten years after light and nimble tractors came on the scene. The most popular of all the OilPulls was the 16-30 Model H.

Advance-Rumely offered a canopy, but many owners opted to make their own or have a custom cab built. Cabs in 1918 were little more than sun shades; it would be a few years before air conditioning and stereos appeared in them.

BELOW: A fuel gauge could be found at any hardware store.

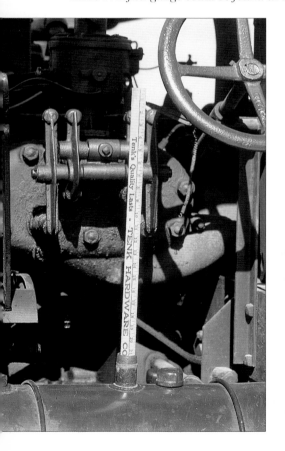

LEFT: OilPulls were built in LaPorte, Indiana by the Advance-Rumely company from 1910 through the late 1920s.

BELOW: The 16-30 was built during the heyday of the gas traction engine and is one of the best of that type of tractor.

RIGHT: Its apparent speed is misleading. OilPull engines operated at a stately 530rpm.

1918 RUMELY OILPULL TRACTOR LA PORTE, IND.

16-30 h.p.

GUARANTEED
TO BURN SUCCESSFULLY ALL GRADES OF KERO-
SENE UNDER ALL CONDITIONS, AT ALL LOADS
UP TO ITS RATED BRAKE HORSE POWER.

ADVANCE RUMELY

ABOVE: With their chimney-like cooling tower puffing out black kerosene smoke, OilPulls are often mistaken for steam engines.

The OilPull used a clever, self-regulating fanless cooling system. Exhaust pumped into the top of the cooling tower created an upward draft, drawing cool air in the bottom. Coolant from the engine was pumped through a radiator located in the tower and cooled by the fresh air. Under heavy load, more heat and exhaust were produced. Additional exhaust increased air flow over the radiator, automatically compensating for the increased heat.

1913
Little Bull

The tractors of the mid-1910s typically weighed 20,000 pounds and required a half-acre field just to turn around in. They were expensive and complicated, not to mention unreliable. While tractor manufacturers were busily building their 10-ton behemoths, farmers clamored for a tractor that was small enough for the average American farm. A man by the name of D.M. Hartsough heard their pleas and designed the Bull tractor. The Bull went on sale in 1913 for $335 dollars, and sold so quickly the factory could not keep up with orders. But with one-wheel-drive and only five drawbar horsepower, the Bull failed to live up to expectations and within two years Bull was effectively out of the tractor business. The Bull experience soured many a small farmer on the small tractor concept, and it took Henry Ford himself to regain their confidence.

SPECIFICATIONS

Weight (pounds): *3200*

Transmission: *1-speed*

Engine (displacement, rpm): *230ci, 750*

Horsepower (belt, drawbar): *12, 5*

Wheels and tires (front, rear):
Single steel front, two steel rear

Years produced: *1913–1915*

Numbers built: *c.5000*

Price new: *$335*

Owner of machine pictured:
Leroy Theilman

ABOVE: A long leather link belt powered the fan. An advanced feature of the Bull was its automotive-type radiator.

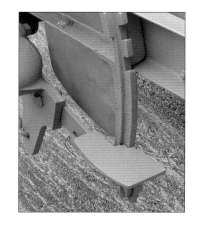

LEFT: The idler wheel on the left side of the tractor could be adjusted to keep the machine level when the drive wheel was in the furrow.

Early Little Bulls had a large arrow atop a cannon ball on the front. The ball added weight to the steer wheel. The arrow told the driver which way the wheel was pointing.

RIGHT: The simple gear train was totally exposed, leading to rapid wear of the gears and bearings.

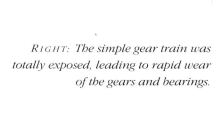

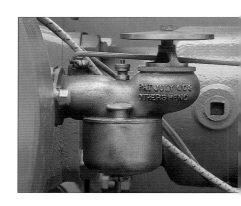

ABOVE: A simple Kingston carburetor fed fuel to the 230ci, 750rpm engine.

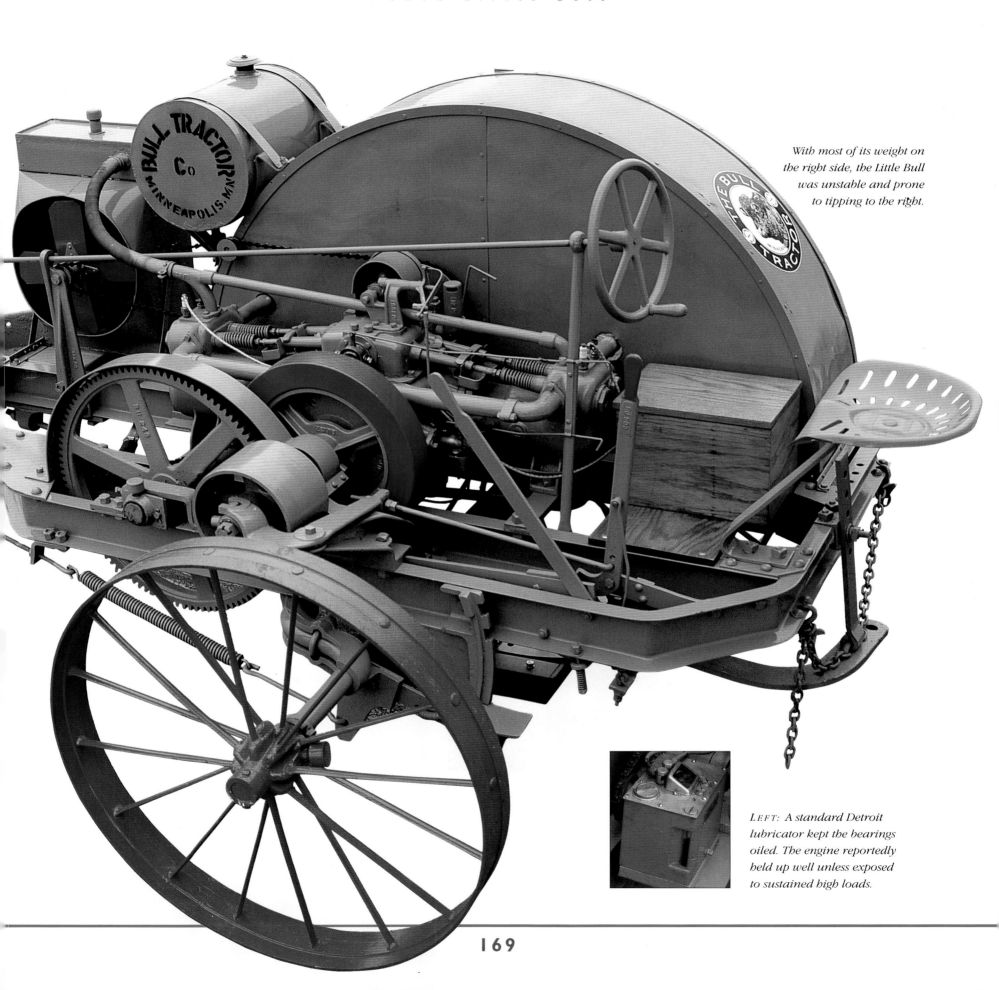

With most of its weight on the right side, the Little Bull was unstable and prone to tipping to the right.

LEFT: A standard Detroit lubricator kept the bearings oiled. The engine reportedly held up well unless exposed to sustained high loads.

ABOVE: Cletrac's AG was a popular agricultural model. With its 28hp gasoline engine, it was a good size for many orchardists and smaller wheat operations.

LEFT: The Cletrac BD was powered by a six-cylinder diesel engine producing 41 belt horsepower.

RIGHT: Cletrac's little Model F was designed primarily as an agricultural tractor. A line of cultivators and other special agricultural equipment was available for it. Its high drive sprocket was an industry innovation that was not appreciated for another 40 years.

that it was incapable of pulling even a single plow under any but the best conditions. Other tractors had been expensive failures for farmers, but because of the sheer number of the tractors sold, more farmers felt hoodwinked by Lyons and the Little Bull than any other make. In spite of attempts to build an adequate tractor, the Bull Tractor Company was out of business by 1919. The Little Bull helped establish a decided negative attitude towards the farm tractor industry. Its influence was instrumental in the establishment of the Nebraska tractor testing laboratory. The Nebraska test became the standard for establishing the veracity of claims make by tractor companies about their machines.

Cletrac

Rollin H. and Clarence G. White established the Cleveland Motor Plow Company in 1916 to produce tracked tractors for the farm trade under the Cletrac name. Cletracs were unique in their use of controlled differential steering. The standard method of steering a crawler had been to declutch the track to the inside of the turn and let the other track do all the pulling. This reduced the tractor's pulling power in poor traction conditions. Cletrac's differential steering placed a planetary gearset controlled by a brake at each drive cog. Controlled amounts of power were sent to each track continuously, improving the tractor's pulling ability.

Cletracs were unique in their use of controlled differential steering.

RIGHT: *A model Cletrac HG made by Meccano.*

construction required tractors as big as the manufacturers could make them. Cletrac continued its close association with the farm trade, building small tractors throughout the 1930s, while courting the construction trade with big 100hp machines. The company produced a small diesel, the Model BD with 27/37 horsepower, in 1935. This tractor was popular with large farmers in the West, where diesel economy and crawler traction were advantageous. In 1939 the company produced its first and only wheel tractor, the 11/17hp Cletrac GG. This tractor was sold through Cletrac dealers in the U.S. and Canada, and through Massey-Harris dealers in Canada. The design was sold to the Avery Company of Louisville, Kentucky in 1942, after only 7527 had been built. Nineteen thirty-nine also saw the Model HG crawler introduced to a ready market. With only 11 drawbar horsepower and weighing 2900 pounds, it was the smallest crawler available in the U.S. Small orchardists and specialty crop growers had needed a crawler like the HG for some time, and it was an immediate success.

Cletrac built small numbers of Model R and H crawlers before hitting its stride with the Model W. The 12/20hp W was built from 1918 through 1932. A Model F 9/16hp tractor was built in 1921 and 1922, and a 15/25 was introduced in 1926. Demonstrating his willingness to look beyond conventional thinking, Rollin White began building push-type implements for his tractors. Front-mounted implements became a hallmark of Cletrac farm equipment that endured throughout the life of the company.

With the rapid spread of the automobile, larger crawlers were required for road construction, and crawler manufacturers were glad to oblige. Farmers generally used crawlers of under 30 horsepower, but

During World War II, farm tractor companies without a crawler in their line began looking at Cletrac as a way to fill out their catalog quickly. The Oliver Corporation bought Cletrac in 1944 (see Chapter Five). Oliver continued to build crawlers, first under the Cletrac name, then as Olivers, through 1965.

ABOVE: *Daily maintenance and lubrication of a crawler tractor is especially important. This manual for the Model AG and BG shows critical lubrication points for the exposed track parts. Notice the forward mounting of the swinging drawbar, designed to keep towed equipment from pulling the tractor sideways in turns.*

1947
Oliver/Cletrac HG

The Cletrac HG made the advantages of a crawler tractor available to small farmers. Weighing just a ton and a half and with only a $2000 price tag, it had no real competition in the marketplace. No crawler matched the HG in versatility. It was available from the factory with track widths of 31 to 68 inches and many different grouser types and even rubber tracks. It was the favorite of growers of specialized crops, who modified it with tracks that varied in width from several feet to only a few inches and grouser plates that raised the chassis up to an extra foot off the ground. Special implements made for the HG ranged from cultivators to loaders. When Oliver Corporation bought Cletrac in 1944, they continued the tractor as the HG until 1951, when they updated it and renamed it the Oliver OC-3. It remained in production until 1957.

LEFT: Oliver Corporation bought the Cleveland Tractor Company (Cletrac) in 1944. Oliver continued to build the Cletrac HG without change except for the Oliver name and Oliver green paint until 1951.

ABOVE: The serial number plate identifies this as an HG 68, indicating the tracks are 68 inches apart.

BELOW: Gauge placement was determined by ease of manufacture, not ease of use.

Compact and light, the HG was a farmers' favorite

SPECIFICATIONS

Weight (pounds): *2950*

Transmission: *3-speed*

Engine (displacement, rpm): *113.1ci, 1400*

Horsepower (belt, drawbar): *17.49, 11.09*

Wheels and tires: *Steel or rubber tracks*

Years produced: *1939–1957*

Numbers built: *29,930*

Price new: *$2048*

Owner of machine pictured: *Ken Garber*

A muffler was available as optional equipment, but the little Hercules engine was so quiet few owners opted for it.

The HG was a no-nonsense tractor. The only concessions to operator comfort were the soft grips on the steering handles, the pan seat, and foot pegs attached to the frame.

BELOW: The belt pulley indicates that this tractor was used for farming. HGs have been seen at every imaginable application, from arctic exploration to dragging carcasses through a slaughterhouse.

The 68-inch track gauge was made for farming. A 42-inch gauge track was offered along with a wide variety of grouser types and widths. A few were sold with rubber tracks, but these proved to be too far ahead of their time and were unreliable.

BELOW RIGHT: This Cockshutt 80 is one of the models Cockshutt purchased from Oliver. Under the red and yellow paint it is 100 per cent Oliver 80.

Cockshutt

The predecessor of the Cockshutt Plow Company was established at Brantford, Ontario, Canada, in 1877 by J.G. Cockshutt. The company concentrated on tillage and other farm equipment, avoiding the more complicated steam and gas traction engines. By the 1920s Cockshutt salesmen recognized that if they had their own tractor on hand to use when demonstrating their plows, they might sell a prospect an expensive tractor as well as a plow. The company began selling Hart-Parr tractors in the early 1920s, then sold Allis-Chalmers tractors in 1929. The Allis-Chalmers arrangement proved unsatisfactory, and Cockshutt was soon back with Hart-Parr.

ABOVE: The Deluxe series Cockshutts reversed the traditional colors, making the sheet metal cream and the frame red.

RIGHT: The Model 30 was the first and most popular of all Cockshutt models. Its two-three plow capacity make it ideal for most American farms.

In the early 1940s discussions had taken place on the subject of Cockshutt building its own tractors. By 1945 it was decided, Cockshutt would build a 30hp tractor. The tractor, called the Model 30, was an extraordinary achievement. Styling was breathtaking, the perfect mix of beauty and business. Under the sheet metal, two features that would soon be accepted as essential on virtually all tractors were built in— continuously running power take-off and hydraulics. The concepts were so new that Cockshutt engineers had to invent a name for them, thus the terms "live" power take-off and "live" hydraulics were born.

The Model 30 went into production in 1946. It was sold in Canada as the Cockshutt 30 and painted orange and sold by Canada's farm co-operative as the CCIL 30. In the United States it was sold as the Cockshutt and Gambles Farmcrest, and also repainted pumpkin orange and sold by

The concepts were so new that Cockshutt engineers had to invent a name for them, thus the terms "live" power take-off and "live" hydraulics were born.

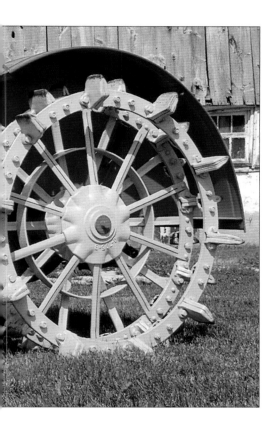

the Farmer's Union Co-Op as the Co-Op E-3. The 30 was powered by a 153ci Buda engine burning gasoline or distillate and had a 6-speed transmission. In 1949 diesel power became available in the form of a 153ci Buda four-cylinder diesel engine. A larger model with a six-cylinder Buda gasoline/distillate engine, heavier 6-speed transmission, and stronger final drive was announced. The Model 40, or Co-Op E-4 in Co-Op's orange livery, offered live PTO in a 30 drawbar, 39 belt horsepower tractor. A Buda diesel engine was made available for the 40 in 1954.

Cockshutt was building and selling tractors about as fast as they could be produced, but recognized a market for both larger and smaller tractors. A 20 drawbar, 26 belt horsepower tractor, the Model 20, was introduced in early 1952. This tractor did not have Cockshutt's famous live power take-off. It was painted orange and sold by Co-Op dealers as the E-2. Later that year a five-plow Model 50 was announced. The 50 and E-5 tractor was similar to the Cockshutt 40/E-4 tractor, but was powered by a 273ci six-cylinder engine. A diesel model was also available. All four models were available with an adjustable wide front axle. The 20, 30, and 40 were also available as tricycle-type row-crop and the 30, 40, and 50 were available as standard-tread tractors.

Rival tractor manufacturer Allis-Chalmers bought Buda, the source of Cockshutt's engines, in 1954, necessitating a change to other brands of engines. A number of variants were made for short periods in the mid-1950s with Hercules or Perkins engines.

Five all-new models were announced in 1958. The tractors wore modern, squared-off sheet metal designed by Raymond Loewy. Under the sheet metal, the smallest tractor, the 540, was powered by a Continental gasoline engine producing 26 drawbar and 31 belt horsepower. It came only as a utility type tractor with live PTO and a three-point hitch with draft control.

The 550 was a row-crop or standard-tread tractor with 34 belt and 26 drawbar horsepower. The 550 could be equipped with a gasoline or diesel engine and a 6-speed transmission. A 560 using the same gearbox and final drive as the 550, was powered by a four-cylinder Perkins engine producing 35/43 horsepower. The largest tractor, the 570, was powered by a 298ci Hercules diesel and produced 40 drawbar horsepower. In 1961 the tractor was given a larger, 339ci engine and renamed the 570 Super.

The 570 was the last unique Cockshutt tractor. In the late 1950s corporate raiders had purchased a controlling interest in Cockshutt and had decided to capitalize on its respected corporate name by selling Fiat tractors under the Cockshutt banner. Then, in January 1962, White Motor Company bought Cockshutt and forced a wedding between Cockshutt and Oliver. From that time on, there were no real Cockshutt tractors. White attempted to wring the last ounce of value from the Cockshutt name by marketing Oliver tractors in Canada under Cockshutt's name and colors. Only the name and the paint were changed, the tractors even used Oliver model numbers. The charade finally ended in the early 1970s, when the Cockshutt name on Oliver tractors was gradually replaced by White nameplates.

ABOVE: With over 44 horsepower, the Cockshutt 50 was a full five-plow tractor. Gasoline fuel Model 50s were powered by 273 cubic inch Buda engines. A 273ci Buda diesel was a popular option, especially on the Wheatland models.

1950
Cockshutt 30

The Cockshutt company of Canada had been in the farm equipment business since 1877, and had become a major supplier to the Canadian farm market. In 1928 they began selling tractors built by other manufacturers, but in 1946 the company took the plunge into tractor manufacturing. Their first tractor, the Model 30, was a remarkably advanced machine. It introduced to the industry live power take-off and live hydraulics, by which the power take-off shaft and hydraulic pump could be operated independently from the drive wheels. Arrangements were quickly made to sell the tractor in the United States as the Gambles Farmcrest, and the Co-Op model E-3. Cockshutt soon bought distributorships and marketed the tractor directly in the United States. The Model 30 not only put Cockshutt in the tractor business, it introduced the company to the lucrative U.S. market and, in so doing, introduced U.S. farmers to the advances of live PTO and hydraulics.

BELOW: Cockshutt engineers tested engines from Continental, Chrysler, and Waukesha before choosing the 153ci, four-cylinder Buda to power the Model 30.

SPECIFICATIONS	
Weight (pounds):	*3434*
Transmission:	*4/6-speed*
Engine (displacement, rpm):	*153ci, 1650*
Horsepower (belt, drawbar):	*28.01, 21.32*
Wheels and tires (front, rear):	*5.50x16, 11.00x38*
Years produced:	*1946–1958*
Numbers built:	*37,328*
Price new:	*$1751*
Owner of machine pictured:	*Henry Shriver*

RIGHT: The operator's station was comfortable by the standards of the 1940s, and offered a clear view down both sides of the tractor, essential for cultivating row crops.

There is an old engineering adage, "If it looks right, it is right." With smoothly integrated sheet metal on a light and nimble-looking frame, the Cockshutt 30 looked just right. Time proved that, in fact, it was just the right tractor at just the right time.

RIGHT: Cockshutt's corporate symbol was an arrow, and when it was time to create graphics for the tractor, a stylized arrow was incorporated into the design. Much of the 30's look is the work of architect Charles Brooks.

LEFT: The words "Made in Canada" are proudly and prominently displayed on the front grille.

BELOW: One effect of the Case buy-out was to improve the looks of Steiger tractors. Traditional Steiger power was still there under the new paint and sheet metal.

Steiger

Steiger didn't build many tractors, but the Steiger brothers came along with a good idea at just the right time. In the mid-1950s farmers were clamoring for bigger tractors with more power. While established tractor manufacturers struggled to build more power into conventional tractors, Douglas and Maurice Steiger looked at the problem from a fresh perspective. They threw out conventional ideas of what a farm tractor ought to be, gathered some industrial truck parts and a big diesel engine, and built themselves a four-wheel-drive articulated tractor. Soon they were building similar tractors for local customers. Major manufacturers immediately took notice and scrambled to put a similar tractor in their lines. Steiger announced a line of more sophisticated tractors in 1963, the 125hp 1200, 216hp 1700, 265hp 2200, and the 350hp 3300.

In 1969 the Bearcat, Wildcat, and Tiger series were introduced. The Bearcat series was built until 1986 with various Cummins and Caterpillar engines of about 225 horsepower. The Wildcat weighed around 15,000 pounds and various versions produced between 175 and 210 horsepower. It was available as both a standard and a row-crop tractor. This model had a 10-speed transmission and lasted until 1980. The largest tractor, the Tiger, was powered by a Cummins V-8 of 320hp through 1976. Steiger unleashed the Tiger III ST450 in 1978. With its 1099ci Caterpillar turbocharged engine it produced 450hp. The next year a 470hp version was released. These two remained in the line until 1984, when the Tiger IV KP 525 replaced them. It was powered by a 1150ci turbocharged Cummins six-cylinder producing 525 horsepower and had a 24-speed transmission.

The Cougar series of 300 engine horsepower tractors came along in 1971 and lasted through the Cougar I and Cougar II models. Less powerful engines were installed in the later models, which remained in the line until 1989. Most came with a 20-speed transmission, but an automatic transmission was available in some models.

The first of the Panther series was announced in 1974. The series lasted until 1989 and included models with 350 to 400 horsepower. Later models featured a

LEFT: Though Steiger tractors were built by farmers for farmers, construction companies were attracted by the high power and relatively low price. Many found their way to construction sites.

"The Caterpillar D2 Diesel Engine—Outstanding features that provide dependable, economical operation."

– Caterpillar sales brochure for the D2

RIGHT: Caterpillar built its own big wheeled tractor, but didn't really compete with the likes of this Cat-powered Steiger.

hydrostatic power take-off with the power output electronically limited to 125 horsepower to avoid damaging PTO equipment. Ten-speed, 20-speed, 12-speed powershift, and 10-speed full automatic transmissions were available on various models over the years.

The agricultural crisis of the early 1980s hit Steiger hard. Sales were poor and the company was deeply in debt. In 1987 Tenneco, the parent company of Case-IH, purchased Steiger. The plant continued to built Steiger models as well as tractors for Case-IH.

When the Steiger brothers built their first articulated tractor, no major manufacturer was building anything remotely similar. But the two farmers from Minnesota knew what large acreage farmers needed and bravely struck out to provide it. By the time the company succumbed to economic pressure, every manufacturer had at least one four-wheel-drive articulated tractor in its catalog.

Caterpillar

Two rivals for the crawler tractor market, the Holt Manufacturing Company and the C.L. Best Gas Tractor Company, buried the hatchet in early 1925 and created the Caterpillar Tractor Company. Holt and Best had battled over the patent rights to crawler tracks for over a decade in lawsuits that had nearly ruined both companies. The new line took on the superior Model 30 and Model 60 tractors from Best's catalog, and the 2-Ton from Holt.

The 2-Ton used an advanced overhead cam engine producing 15 drawbar and 25 belt horsepower. From there it was a big jump to the 30, which weighed 8700 pounds and produced 25 drawbar horsepower. The 60 weighed 19,095 pounds and produced 50 drawbar horsepower. While they were developed for agriculture, logging and road construction companies soon became Caterpillar's biggest customers.

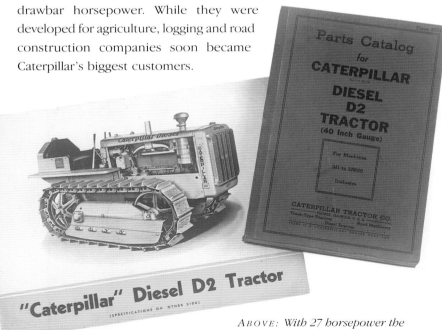

LEFT: Best had a such a successful tractor in the Model 30 that it remained in the line after Caterpillar was formed.

ABOVE: With 27 horsepower the D2, introduced in 1939, was a suitable size for farm use. An orchard version with no projections from the hood and fenders that covered the tracks was also available.

The success of the 30 and 60 helped make "Caterpillar" a generic term for any crawler tractor.

RIGHT: Larger crawlers were appropriate for construction, but as these brochures show, the Caterpillar D2 was intended for farm use.

BELOW: Like the tractor, this brochure for the Caterpillar 30 is a sought-after collector's item.

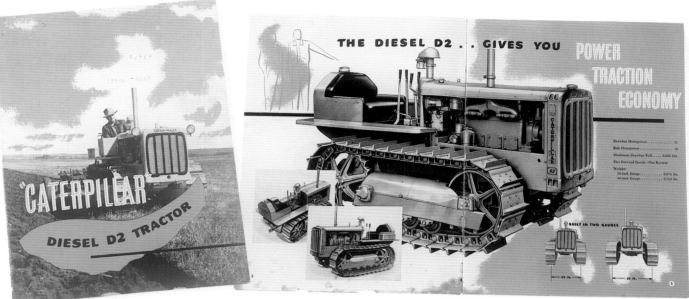

RIGHT: The 60 was the largest tractor in the Caterpillar line. Many an American road was built with Cat 60 power.

BELOW LEFT: The success of the 30 and 60 helped make "Caterpillar" a generic term for any crawler tractor.

Over the decades, the tractors were designed around the needs of its principal customers. Some efforts were made to adapt the smaller models to agriculture, but farmers who needed crawler tractors largely made do with what was available to the construction trade.

All that changed in the early 1980s, when Caterpillar presented large acreage farmers with an entirely new tractor. It had high power to weight ratio, its weight was distributed forward, where it should be for drawn implements, and it floated on air-suspended rubber tracks. Caterpillar called the new machine the Challenger. The first model, announced in 1986, was the 65. It weighed 31,000 pounds, had 270 engine horsepower, a 10-speed powershift transmission, and cost over

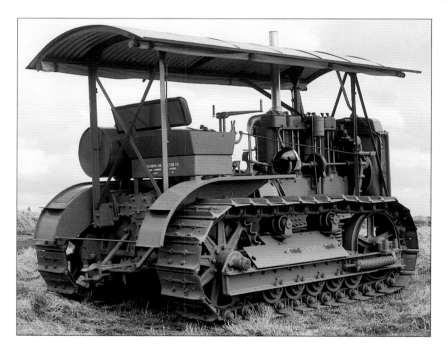

$140,000. In comparison, a 270hp articulated four-wheel-drive tractor weighing 23,000 pounds cost about $110,000 and a 270hp crawler built for the construction trade weighed 60,000 pounds and cost a staggering $282,590. The Challenger was certainly a challenge to the conventional tracked tractor, but was still significantly heavier and more costly than a wheel tractor.

RIGHT: This spring-powered tin toy Caterpillar is a coveted collectible today.

BELOW: The first Challenger was the 15-ton Model 65. Farmers soon began asking for a smaller version.

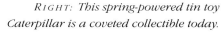

In 1990 an even larger model, the 75 with 325 horsepower, was announced. Both the 65 and 75 continued through the end of the decade with periodic power upgrades. Nineteen ninety-three saw the introduction of the 85 series, with 355 engine horsepower.

Significant additions to the line were made in 1995. Two rubber-tracked row-crop tractors were announced, the 35 and the 45. The undercarriages of these carried the chassis higher for more ground clearance, and the tracks could be adjusted to provide from 60 to 88 inches between their centers. The tracks themselves were available in widths of 16 to 32 inches. The 35 produced 210 engine horsepower, while the 45 produced 240. The new row-crops came with an electronically controlled 16-speed powershift transmission that allowed sequential shifting one gear at a time, automatic sequential shifting, shifting to a preselected gear, programmed shifting, or automatic shifting through only the upper six speeds. A 265hp version, the Model 55, was introduced in 1996. At the top end of the scale, a big 410hp Model 95E was introduced in 1999.

The 95E was the last new Caterpillar of the century, but the Challenger series had left its mark. Other tractor manufacturers were scrambling to produce similar machines but, in this type of tractor, Caterpillar led the field into the new century.

ABOVE: The R2 shared all the features of the D2, but offered them in a gasoline-powered tractor.

ABOVE: Powered by a three-cylinder Caterpillar-built diesel engine, the RD6 was the successor to the Diesel 40.

RIGHT: The Challenger 35 and 45 brought a "revolutionary change for row-crop tractors."

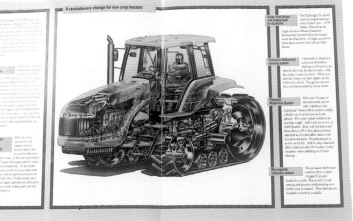

BELOW: The 6195, the largest of White's 6100 series, was only available with front-wheel-assist.

White

White Motor Corp. had purchased the Oliver Corporation in 1960, Cockshutt Farm Equipment in 1962, and Minneapolis-Moline in 1963 in addition to numerous smaller companies to form, in 1969, the White Farm Equipment Company. White mixed and matched components from Minneapolis-Moline and Oliver and added imported tractors to provide three nearly identical lines of tractors, which were sold under the Oliver, Cockshutt, and Minneapolis-Moline names.

In 1974 White dropped the pretense that its Oliver, Cockshutt, and Minneapolis-Moline lines were different tractors, and consolidated them all under the White name. White brand tractors were painted silver and given distinctive, squared-off sheet metal. The first new tractor to be announced was the Field Boss 4-150, an articulated tractor powered by a 175hp Caterpillar engine. White quickly moved into the popular two-wheel-drive market with the 2-105 and 2-150. The 2-105 was a new 106 PTO

RIGHT: White marketed the Minneapolis-Moline A4T 1600 in Canada as the Plainsman. It was sold in the United States as both the Minneapolis-Moline A4T and the Oliver 2655. The tractors were powered by a Minneapolis-Moline 585ci diesel producing169 horsepower.

In 1974 White dropped the pretense that its Oliver, Cockshutt, and Minneapolis-Moline lines were different tractors, and consolidated them all under the White name.

RIGHT: White's American series was available in Oliver green, Cockshutt red, Minneapolis-Moline gold, or White silver. The color choices tried to capitalize on the reputations these names had built in the tractor business.

BELOW: The White Field Boss 4-150 combined two Oliver two-wheel-drive power trains hinged in the middle, and a Cummins V-8 diesel. The result was a unique low profile articulated tractor with 175 engine horsepower and sliding-axle wheel tread adjustment.

painted Oliver colors, but after 1976 painted in White silver. White's largest two-wheel-drive tractor came along in 1977. The 2-180 was powered by a 636ci Caterpillar V-8 diesel. The two Fiat-built tractors were dropped and the small end of the line, comprised of four utility tractors between 28 and 61 horsepower, was filled with Iseki tractors imported from Japan.

White proved to be more adept at finishing off tractor companies than growing them. After extinguishing some of the brightest stars in American agriculture, White itself failed. In 1980 White Farm Equipment was sold to an investment company. The new owner put the company back on its feet temporarily, and continue building the White tractors under the WFE name.

The lower end of the WFE line was filled with five sizes of Iseki tractors from 28 to 75 horsepower. Otherwise the tractor line saw few changes until 1984 when two big four-wheel-drives, the 225hp 4-225, and the 270hp 4-270 were announced.

White Farm Equipment's new owners were even less successful than the previous landlords, and in 1985 filed for bankruptcy. The White line was sold again. The new owners had already purchased the New Idea farm equipment company, and the new company was called White-New Idea. The tractor line changed little as the company tried to get its legs under it. In 1986 sweeping changes were made in the domestically built tractors. Engines for all models were purchased from CDC, the engine builder created by a joint venture between Case and Cummins. Five tractors ranging from 94 to 188 horsepower were announced. All had an 18-speed transmission, with front-wheel-assist available as an option. In 1989 the American 60 and 80 were added. With 61 and 81 horsepower, these were among the smallest tractors sold in America that were also built in America. They came standard with a 6-speed transmission, with a 3-speed power shift offered as an option.

One final series was offered under the White name. These tractors were essentially the same as the previous line, with the 94hp Model 100 dropped. The "new" tractors were called the White Workhorse line. In 1993 White tractors got another new owner when AGCO purchased the tractor line. AGCO assembled a line of large tractors from White and Deutz parts, filling the smaller sizes with SAME tractors imported from Italy.

horsepower tractor. The 2-150, with 148 horsepower, was based on the Minneapolis-Moline 1355. The next year a larger articulated tractor with a 210hp engine was announced. White continued to add to the lower end of the line with the 2-85, a 2-105 with a detuned engine giving 86hp.

Two big two-wheel-drive row-crop tractors were added in 1976. These shared the Oliver 18-speed transmission and the 478ci turbocharged Hercules engine that had been used in the Oliver 2150. The 2-135 was rated at 138 horsepower and the 2-155 produced 158 horsepower. In 1978 the option of front-wheel-assist was offered on this pair. Also in 1976 a small row-crop tractor based on the Oliver 1655, the 2-70, was announced. The 2-70 was available with a 70hp gasoline or diesel engine and front-wheel-assist. Two utility tractors also came along in 1976, the 47hp 2-50 and the 63hp 2-60. These two were Fiat tractors, initially imported and

AGCO has a tractor that's right for any farmer, anywhere in the world.

AGCO

America's newest tractor manufacturer, AGCO, was formed in 1990 when American management personnel of Deutz-Allis, which had purchased the Allis-Chalmers Agricultural Equipment Company in 1985, formed the Allis-Gleaner Corporation (AGCO). AGCO then purchased Deutz-Allis from its parent company, Klockner-Humbolt-Deutz AG of Cologne, Germany and began marketing the Deutz-Allis line under the AGCO-Allis brand name. In 1993 AGCO purchased Massey-Ferguson's North American operations and the struggling White-New Idea line. The following year, AGCO purchased the remainder of Massey-Ferguson's world-wide operation. In 1995 AGCO purchased the McConnell Tractors Ltd of Brantford, Ontario. McConnell had been building four-wheel-drive articulated tractors based on old Massey-Ferguson designs.

ACGO's strategy has been the same as White's was initially, to market several brands of tractors that share varying amounts of parts in common. Three major lines are marketed by AGCO in North America, AGCO-Allis, AGCO-White, and Massey-Ferguson. A line of articulated four-wheel drive tractors was marketed under the AGCOSTAR name.

ABOVE: Though heavily influenced by European design, AGCO-Allis tractors offered increased American content as the line matured. With Allis, White, Massey-Ferguson, and AGCOSTAR brands all under the AGCO banner, AGCO has a tractor that's right for any farmer, anywhere in the world.

ABOVE: Unveiled in 1994, the 9435 and 9455 introduced ESC, a new dimension in electronic tractor control and systems monitoring.

Initially, Deutz-Allis large tractors were given a new electronically controlled transmission and marketed under the AGCO-Allis name. The old line of small tractors was dropped and all tractors of under 120hp were purchased from SAME-Lamborghini-Hurliman (S+L+H). In 1993, tractors in the 135 to 195hp range were powered by Deutz engines coupled to a new electronically controlled 18-speed power shift transmission built by White. The next year Detroit Diesel liquid-cooled engines were offered as well as air-cooled Deutz engines and a 215hp model was added. The Deutz engines were phased out by 1996. In 1997 the AGCO-Allis and AGCO-White lines were completely redesigned, though along parallel lines. At the end of the century, all AGCO-Allis tractors between 70 and 145 horsepower were manufactured in Massey-Ferguson plants in Europe. Smaller tractors were built in England, while the 160 through 225hp tractors were built in the U.S.

A similar program was followed for AGCO-White. In 1992 the small Japanese tractors were dropped and all tractors under 120 horsepower were purchased from S+L+H. Cummins-powered machines from 135 to 231 horsepower and built in the United States made up the rest of the line. In 1997 things changed drastically, with six tractors in the 70 to 145hp range being built in European Massey-Ferguson plants with engines supplied by Cummins. Four U.S.-built tractors were added, the 160hp 8510, the 180hp 8610, the 200hp 8710, and the largest AGCO-White model, the 225hp 8810. All U.S.-built models featured an 18-speed powershift transmission shared by AGCO-Allis tractors and offered front-wheel-assist. They were all powered by the same basic Cummins engine in different states of tune and used many components in common with AGCO-Allis tractors.

BELOW: AGCO has followed a strategy of offering parallel lines of similar tractors using many common components. These three AGCO tractors are more similar than they may appear.

In 1995 AGCO purchased the McConnell Tractors Ltd, moved the operation to its Coldwater, Ohio plant and began building a modified version of the McConnell. Called AGCOSTAR, the tractor came in two sizes. The 360hp 8360 was powered by an 855ci Cummins six-cylinder diesel. The 8425, with 425 horsepower, used the same engine in a higher state of tune. Both tractors had a partially synchronized 18-speed transmission.

AGCO ended the century with four lines of farm tractors in its catalog. AGCO-Allis, AGCO-White, Massey-Ferguson, and AGCOSTAR. AGCO-Allis and AGCO-White were mostly parallel lines with many models sharing power trains. The Massey-Ferguson branch offered a more extensive line of tractors, particularly in the small tractor range, with many unique models. AGCOSTAR consisted of the two big articulated tractors. Though it is a company without a heritage selling to a history-conscious market, with its global reach (AGCO owned many smaller overseas tractor companies) and its deep reach into the North American market, AGCO ended the century in a good position to be still around for the next centennial celebration.

ABOVE: The rounded, compact look of the year 2000 tractors is a refreshing change after the bulky squared hoods that were the fashion in the 1970s.

COMPARISON TABLES

Allis-Chalmers

How does the 1938 Allis-Chalmers WC stack up?
A comparison of the Allis-Chalmers WC and similar contemporary tractors

Make	price	rated hp belt/drawbar	weight pounds	engine capacity	trans. speeds
Allis-Chalmers WC	$992	27/18	3325	201ci	4
Case CC	$1174	29/23	4090	260ci	3
Farmall F-20	$1035	22/13	3950	221ci	4
John Deere A	$1095	30/26	4059	309ci	4
Massey-Harris Challenger	$1205	35/30	3915	248ci	4
Minneapolis-Moline Z	$1131	31/26	3750	186ci	5
Oliver 70	$1045	28/22	3100	201ci	4

* The figures quoted in this table apply to tractors fitted with rubber tires

How does the 1963 Allis-Chalmers D-19 diesel stack up?
A comparison of the Allis-Chalmers D-19 and similar contemporary tractors

Make	price	rated hp belt/drawbar	weight pounds	engine capacity	trans. speeds
Allis-Chalmers D-19	$5972	67/62	6835	262ci	8
John Deere 3010	$4335	59/53	5606	254ci	8
Massey-Ferguson S-90	$5722	68/60	5726	242ci	8
Minneapolis-Moline M5	$5512	58/51	6965	336ci	10
Oliver 1600	$6163	70/62	7180	265ci	6

How does the 1974 Allis-Chalmers 7080 stack up?
A comparison of the Allis-Chalmers 7080 and similar contemporary tractors

Make	price	rated hp belt/drawbar	weight pounds	engine capacity	trans. speeds
Allis-Chalmers 7080	$54,560	182/154	14,605	426ci	20
Case 1570	$30,075	180/153	16,290	504ci	12
John Deere 4840	$57,650	181/151	18,130	466ci	8
Massey-Ferguson 2805	$60,200	194/158	15,630	640ci	24

Case

How does the 1919 Case 15-27 stack up?
A comparison of Case 15-27 and similar contemporary tractors

Make	price	rated hp belt/drawbar	weight pounds	engine capacity	trans. speeds
Case 15-27	$1350	27/15	6400	382ci	2
Allis-Chalmers 15-30	$2100	30/15	6000	460ci	2
Keck-Gonnerman	$1500	30/15	6500	706ci	2
Lauson	$2150	30/15	5750	425ci	2
Rumely OilPull	$2400	30/16	9600	654ci	2

How does the 1938 Case Model L stack up?
A comparison of Case Model L and similar contemporary tractors

Make	price	rated hp belt/drawbar	weight pounds	engine capacity	trans. speeds
Case L	$1540	47/41	5736	403ci	3
Allis-Chalmers A	$1495	51/40	7120	461ci	4
Huber Super 4	$2235	70/50	9450	617ci	2
McCormick-Deering W40	$1587	50/35	6695	298ci	4
Oliver 99	$2157	62/52	6467	443ci	4

How does the 1962 Case 730 diesel stack up?
A comparison of Case 730 diesel and similar contemporary tractors

Make	price	rated hp belt/drawbar	weight pounds	engine capacity	trans. speeds
Case 730	$5008	56/51	6095	251ci	8
Cockshutt 570	$4700	54/40	6320	298ci	6
John Deere 3010	$4335	59/53	5606	254ci	8
Farmall 460	$3440	51/47	4419	220ci	5
Minneapolis-Moline M5	$5512	58/51	6965	336ci	10

How does the 1978 Case 2870 diesel stack up?
A comparison of Case 2870 diesel and similar contemporary tractors

Make	price	rated hp belt/drawbar	weight pounds	engine capacity	trans. speeds
Case 2870	$56,970	252/210	25,100	674ci	12
Allis-Chalmers 8550	$85,750	254/218	27,040	731ci	20
Belarus 7100	$70,000	270/250	32,460	1360ci	16
Versatile 875	$109,395	247/218	25,320	855ci	12

Ferguson

How does the 1958 Massey-Ferguson 65 stack up?
A comparison of the Massey-Ferguson 65 and similar contemporary tractors

Make	price	rated hp belt/drawbar	weight pounds	engine capacity	trans. speeds*
Massey-Ferguson 65	$3935	41/33	3843	176ci	6
Ford 851	$3187	43/32	3187	172ci	5
IHC 350U	$3750	38/31	4285	175ci	10/2
Minneapolis-Moline 445U	$2750	38/30	4240	206ci	10
Oliver 550	$2918	41/35	3655	155ci	6

* second figure indicates ranges

How does the 1968 Massey-Ferguson MF1130 stack up?
A comparison of the Massey-Ferguson MF1130 and similar contemporary tractors

Make	price	PTO hp	weight pounds	engine capacity	trans. speeds*
Massey-Ferguson MF1130	$12,425	121	11,500	354ci=	12/4
Allis-Chalmers 220	$13,435	135	10,624	426ci=	8/2
Case 1070	$26,135	107	11,100	451ci	12/4
Farmall 1256	$10,290	116	10,300	407ci=	16/8
John Deere 4520	$11,725	122	12,390	404ci=	8/2
Oliver 2050	$12,545	119	14,000	478ci	18/3
Minneapolis-Moline G1000	$9,650	111	10,400	504ci	10/2

* second figure indicates ranges   = turbocharged

How does the 1985 Massey-Ferguson 240 stack up?
A comparison of the Massey-Ferguson 240 and similar contemporary tractors

Make	price	PTO hp	weight *pounds*	engine capacity	trans. speeds*
Massey-Ferguson 240	$19,165	34	4015	152ci	8/2
Case-IH 385	$16,745	35	4920	155ci	8/4
Deutz 3607	$11,364	33	4120	115ci	8/2
Ford 2110	$11,685	35	3435	139ci	12/4
John Deere 1050	$13,390	33	2800	105ci	16/8
Yanmar YM330	$8165	33	2552	91ci	8/2

* second figure indicates ranges

Ford

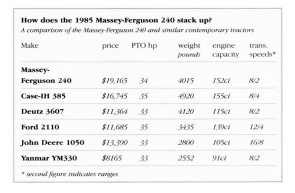

How does the 1928 Fordson F stack up?
A comparison of the Fordson F and similar contemporary tractors

Make	price	rated hp *belt/drawbar*	weight *pounds*	engine capacity	trans. speeds
Fordson F	$385	21/11	2710	251ci	3
John Deere GP	na	20/10	3600	312ci	3
McCormick-Deering 10-20	$875	20/10	3700	284ci	3
Wallis 12-20	$950	20/12	3544	248ci	3

How does the 1939 Ford 9N stack up?
A comparison of the Ford 9N and similar contemporary tractors

Make	price	rated hp *belt/drawbar*	weight *pounds*	engine capacity	trans. speeds
Ford 9N	$585	20/13	2340	120ci	3
Allis-Chalmers C	$1060	20/14	3030	125ci	4
Farmall A*	$565	16/13	3395	113ci	4
John Deere B	$867	17/14	3725	175ci	6
Oliver 60	$1117	17/13	3100	121ci	4

*data are for a Farmall A using gasoline fuel

How does the 1956 Ford 960 stack up?
A comparison of the Ford 960 and similar contemporary tractors

Make	price	rated hp *belt/drawbar*	weight *pounds*	engine capacity	trans. speeds
Ford 960	$2380	40/30	3280	172ci	5
Allis-Chalmers WD-45	$2915	38/30	3780	226ci	4
John Deere 620	$3740	41/33	6150	303ci	6
Minneapolis-Moline 445	$3050	38/31	3900	206ci	10

How does the 1968 Ford 8000 stack up?
A comparison of the Ford 8000 and similar contemporary tractors

Make	price	PTO hp	weight *pounds*	engine capacity	trans. speeds
Ford 8000	$9060	106	9300	401ci	16
Case 1070	$20,400	107	11,100	451ci	12
Farmall 856	$10,390	100	9530	301ci	16
John Deere 4320	$10,700	100	10,242	404ci	16
Oliver 1950T	$12,855	105	10,900	310ci	18

How does the 1979 Ford 9700 stack up?
A comparison of the Ford 9700 and similar contemporary tractors

Make	price	PTO hp	weight *pounds*	engine capacity	trans. speeds
Ford 9700	$24,165	136	11,458	401ci	8
Case 1270	$29,690	135	12,400	451ci	12
John Deere 4440	$45,145	130	10,900	466ci	16
Massey-Ferguson 2745	$44,360	144	13,400	539ci	24
White 2-135	$41,740	138	14,015	478ci	18

Hart-Parr/Oliver

How does the 1921 Hart-Parr 15-30 stack up?
A comparison of the Hart-Parr 15-30 and similar contemporary tractors

Make	price	rated hp *belt/drawbar*	weight *pounds*	engine capacity	trans. speeds
Hart-Parr 30	$1160	30/15	5570	465ci	2
Allis-Chalmers 15-30	$2100	30/15	6000	460ci	2
Case 15-27	$1350	27/15	6400	382ci	2
McCormick-Deering 15-30	$1250	30/15	5750	382ci	3
Lauson	$2150	30/15	5750	425ci	2
Rumely OilPull	$2400	30/16	9600	654ci	2

How does the 1930 Oliver Hart-Parr 18-28 stack up?
A comparison of the Oliver Hart-Parr 18-28 and similar contemporary tractors

Make	price	rated hp *belt/drawbar*	weight *pounds*	engine capacity	trans. speeds
Oliver Hart-Parr 18-28	$1025	28/18	4000	281ci	3
Allis-Chalmers U	$895	30/19	4125	301ci	4
Case C	$950	27/17	4105	260ci	3
Rock Island G-2	$1045	30/18	4200	346ci	2
Twin City 17-28	$1100	28/17	5050	341ci	2
Wallis 20-30	$945	30/20	4096	346ci	2

How does the 1940 Oliver 60 stack up?
A comparison of the Oliver 60 and similar contemporary tractors

Make	price	rated hp *belt/drawbar*	weight *pounds*	engine capacity	trans. speeds
Oliver 60	$1117	17/13	3100	121ci	4
Allis-Chalmers C	$1060	20/14	3030	125ci	3
Case RC	$875	17/11	3700	133ci	3
Farmall A	$575	15/12	3395	113ci	4
John Deere B	$867	17/14	3725	175ci	6

How does the 1952 Oliver 77 stack up?
A comparison of the Oliver 77 and similar contemporary tractors

Make	price	rated hp *belt/drawbar*	weight *pounds*	engine capacity	trans. speeds
Oliver 77	$2261	33/26	4032	194ci	6
Allis-Chalmers WD	$1930	30/24	3975	201ci	4
Case DC	$2165	33/26	6835	260ci	4
Farmall H	$1898	23/19	5375	152ci	5
John Deere A	$2297	34/26	4909	321ci	6

How does the 1967 Oliver 1950T stack up?
A comparison of the Oliver 1950T and similar contemporary tractors

Make	price	PTO hp	weight *pounds*	engine capacity	trans. speeds*
Oliver 1950T	$10,455	105	9900	310ci	18/6
Farmall 856	$10,390	100	9530	301ci	16/8
Case 1070	$20,400	107	11,100	451ci	12/4
Ford 8000	$9060	106	9300	401ci	16/4
John Deere 4320	$10,700	100	10,242	404ci	16/6

* second figure indicates ranges

International Harvester

How does the 1928 McCormick-Deering 10-20 stack up?
A comparison of the McCormick-Deering 10-20 and similar contemporary tractors

Make	price	rated hp *belt/drawbar*	weight *pounds*	engine capacity	trans. speeds
McCormick-Deering 10-20	$875	20/10	3700	284ci	3
Fordson F	$385	21/11	2710	251ci	3
John Deere GP	na	20/10	3600	312ci	3

How does the 1938 Farmall F-20 stack up?

A comparison of the Farmall F-20 and similar contemporary tractors

Make	price	rated hp belt/drawbar	weight pounds	engine capacity	trans. speeds
Farmall F-20	$875	23/15	3950	221ci	3
Allis-Chalmers WC	$992	30/22	2700	201ci	4
Case CC	$975	29/23	4090	260ci	3
John Deere A	$1095	30/26	4059	309ci	6
Oliver 70	$1045	28/22	3100	201ci	4

How does the 1950 Farmall M stack up?

A comparison of the Farmall M and similar contemporary tractors

Make	price	rated hp belt/drawbar	weight pounds	engine capacity	trans. speeds
Farmall M	$2346	37/33	4910	248ci	5
Allis-Chalmers WD	$1930	26/24	3975	201ci	4
Case DC	$2165	37/33	6835	260ci	4
John Deere A	$2297	38/34	4909	321ci	6
Massey-Harris 30	$1754	34/26	3770	163ci	5
Minneapolis-Moline ZA	$2025	36/32	3750	206ci	5
Oliver 88	$2660	42/37	5110	231ci	6

How does the 1959 International 660 stack up?

A comparison of the International 660 and similar contemporary tractors

Make	price	rated hp belt/drawbar	weight pounds	engine capacity	trans. sp/rgs
IHC 660	$5980	81/70	7800	263ci	10/2*
Case 900	$5505	70/66	7882	377ci	6
John Deere 830	$5775	73/67	8745	472ci	6
Massey-Ferguson 88	$4325	63/53	6600	277ci	8
Minneapolis-Moline GBD	$5965	63/55	7400	426ci	5
Oliver 990GM	$8070	84/77	8155	213ci	6

(forward/reverse)

How does the 1983 International 3288 stack up?

A comparison of the International 3288 and similar contemporary tractors

Make	price	PTO hp	weight pounds	engine capacity	trans. speeds*
IHC 3288	$40,000	91	10,432	358ci	16/8
Allis-Chalmers 6080	$28,620	84	6915	200ci	12/3
Case 1690	$31,650	90	8710	329ci	12/4
John Deere 4040	$34,550	90	9960	404ci	8/2
Massey-Ferguson 2640	$37,900	91	10,940	354ci	16/12

* second figure indicates ranges

John Deere

How does the 1924 John Deere D stack up?

A comparison of the John Deere D and similar contemporary tractors

Make	price	rated hp belt/drawbar	weight pounds	engine capacity	trans. speeds
John Deere D	$1125	27/15	4403	465ci	2
Case 15-27	$1350	27/15	6400	382ci	2
Keck-Gonnerman	$1500	30/15	6500	706ci	2
Lauson	$2150	30/15	5750	425ci	2
Rumely OilPull H	$2400	30/16	9600	654ci	2

How does the 1938 John Deere G stack up?

A comparison of the John Deere G and similar contemporary tractors

Make	price	rated hp belt/drawbar	weight pounds	engine capacity	trans. speeds
John Deere G	$995	31/21	4488	413ci	4
Farmall F-30	$1100	30/20	5300	284ci	4
Minneapolis-Moline KTA	$950	37/23	4300	284ci	3
Oliver 80	$1095	35/23	4565	334ci	3

How does the 1948 John Deere MT stack up?

A comparison of the John Deere MT and similar contemporary tractors

Make	price	rated hp belt/drawbar	weight pounds	engine capacity	trans. speeds
John Deere M	$1523	18/14	2520	101ci	4
Allis-Chalmers CA	$1505	23/18	2835	125ci	4
Farmall C	$1508	19/15	2710	113ci	4
Massey-Harris 81 R	$1121	24/16	2720	124	4

How does the 1958 John Deere 720 diesel stack up?

A comparison of the John Deere 720 and similar contemporary tractors

Make	price	rated hp belt/drawbar	weight pounds	engine capacity	trans. speeds*
John Deere 720	$3940	50/40	7105	376ci	6
Allis-Chalmers D-17	$4970	45/35	5540	262ci	8/2
Cockshutt 570	$4700	54/40	6320	298ci	6
Farmall 450	$4048	43/35	4720	281ci	10/2
Minneapolis-Moline 5-Star	$4316	49/39	6070	336ci	10/2
Oliver 880	$4516	53/41	5350	265ci	6

* second figure indicates ranges

How does the 1969 John Deere 4520 diesel stack up?

A comparison of the John Deere 4520 and similar contemporary tractors

Make	price	rated PTO hp	weight pounds	engine capacity	trans. speeds*
John Deere 4520	$11,725	122	12,390	404ci	8/2
Allis-Chalmers D-21 II	$12,920	128	10,625	426ci	8/2
Farmall 1256	$10,290	116	10,300	407ci	16/8
Ford 9000	$11,915	131	10,725	401ci	16/8
Minneapolis-Moline G1000 Vista	$10,775	111	10,525	504ci	10/2
Oliver 1950T	$10,455	105	9900	310ci	18/6

* second figure indicates ranges

How does the 1979 John Deere 8640 stack up?

A comparison of the John Deere 8640 and similar contemporary tractors

Make	price	engine hp	weight pounds	engine capacity	trans. speeds*
John Deere 8640	$77,630	275	24,470	619ci	16/4
Case 2870	$56,970	300	21,250	673ci	12/4
International 4586	$68,575	300	22,400	800ci	10/2
Steiger Panther III PTA297	$95,000	295	28,322	903ci	10/2
Versatile 875	$109,395	280	20,106	855ci	12/4

* second figure indicates ranges

How does the 1995 John Deere 7200 stack up?

A comparison of the John Deere 7200 and similar contemporary tractors

Make	price	PTO hp	weight pounds	engine capacity	trans. speeds
John Deere 7200	$39,620	94	11,140	359ci	16
AGCO-Allis 7600	$35,975	90	8686	317ci	24
Belarus 905	$22,125	92	8560	290ci	18
Case-IH 5240	$38,355	100	9800	359ci	16
Ford 8420	$37,965	96	8957	456ci	16

Massey-Harris

How does the 1935 Massey-Harris Model 25 stack up?

A comparison of the Massey-Harris Model 25 and similar contemporary tractors

Make	price	rated hp belt/drawbar	weight pounds	engine capacity	trans. speeds
Massey-Harris 25	$1589	26/41	4917	346ci	3
Allis-Chalmers E	$1395	25/40	7000	511ci	2
Case L.	$1495	26/40	5157	403ci	3
Oliver 28-44	$1325	29/45	5575	443ci	3

How does the 1949 Massey-Harris 22 stack up?
A comparison of the Massey-Harris 22 and similar contemporary tractors

Make	price	rated hp belt/drawbar	weight pounds	engine capacity	trans. speeds
Massey-Harris 22	$1404	27/18	2560	140ci	4
Allis-Chalmers C	$1455	20/14	3030	125ci	3
Farmall C	$1508	19/15	2710	113ci	4
Ford 8N	$1329	22/17	2410	120ci	4
John Deere MT	$1532	18/14	3200	101ci	4

How does the 1954 Massey-Harris Model 44 Special stack up?
A comparison of the Massey-Harris Model 44 Special and similar contemporary tractors

Make	price	rated hp belt/drawbar	weight pounds	engine capacity	trans. speeds
Massey-Harris 44	$2580	43/34	5185	277ci	5
Farmall Super M	$2538	41/33	5725	264ci	5
John Deere 70	$2855	43/33	6035	380ci	6
Minneapolis-Moline UB	$2492	43/34	5700	283ci	5
Oliver 88	$2810	38/29	5110	231ci	6

Minneapolis-Moline

How does the 1932 M-M Twin City MT stack up?
A comparison of the M-M Twin City MT and similar contemporary tractors

Make	price	rated hp belt/drawbar	weight pounds	engine capacity	trans. speeds
M-M MT	$985	25/14	4860	284ci	3
Farmall	$950	22/13	3650	221ci	3
John Deere GP WT	$1050	24/16	4141	339ci	3
Massey-Harris GP	$1050	20/13	3861	226ci	3

How does the 1949 M-M GTB stack up?
A comparison of the M-M GTB and similar contemporary tractors

Make	price	rated hp belt/drawbar	weight pounds	engine capacity	trans. speeds
M-M GTB	$3271	51/39	6600	403ci	5
Case LA	$3030	43/35	6516	403ci	4
IH W-9	$3162	42/33	6215	335ci	5
Massey-Harris 55	$3214	52/41	7150	382ci	4
Oliver 99	$3528	54/41	6966	443ci	4

How does the 1959 M-M 4-Star stack up?
A comparison of the M-M 4-Star and similar contemporary tractors

Make	price	rated hp belt/drawbar	weight pounds	engine capacity	trans. speeds*
M-M 4-Star	$3298	37/30	3750	206ci	10/2
Farmall 460	$3325	44/35	6020	236ci	10/2
John Deere 630	$3328	41/33	6150	303ci	6
Massey-Ferguson 65	$3635	41/33	3802	176ci	6
Oliver 770	$2969	44/34	4947	217ci	6

* second figure indicates ranges

How does the 1969 M-M G1000 Vista stack up?
A comparison of the M-M G1000 Vista and similar contemporary tractors

Make	price	PTO hp	weight pounds	engine capacity	trans. speeds*
M-M G1000 Vista	$10,775	111	10,525	504ci	10/2
Case 1030	$10,390	102	9280	451ci	8/2
Farmall 1256	$10,290	116	10,300	407ci	16/8
Ford 9000	$11,915	131	10,725	401ci	16/8
John Deere 4520	$11,725	122	12,390	404ci	8/2
Oliver 1950T	$10,455	105	9900	310ci	18/6

* second figure indicates ranges

INDEX

Bibliography

150 Years of International Harvester, C.H.
   Wendel
150 Years of J.I. Case, C.H. Wendel
Agricultural Tractor 1855-1950, The, R.B. Gray
Allis Chalmers Story, The, C.H. Wendel
*Allis-Chalmers Farm Equipment 1914-
   1985*, Norm Swinford
American Farm Tractors in the 1960s,
   Chester Peterson Jr. and Rod Beemer
Corporate Tragedy, A, Barbara Marsh
Encyclopedia of American Farm Tractors,
   C.H. Wendel
Farm Tractors 1950-1975, Lester Larsen
Farm Tractors 1975-1995, Larry Gay
Farmall Letter Series Tractors, Guy Fay
   and Andy Kraushaar
Ford and Fordson Tractors, Michael
   Williamson
Ford Tractor Story, The, Stuart Gibbard
Ford: Decline and Rebirth, Allan Nevins
   and Frank Ernest Hill
Ford: Expansion and Challenge, Allan
   Nevins and Frank Ernest Hill
Full Steam Ahead, David Erb and Eldon
   Brumbaugh

Global Corporation, A, E.F. Neufeld
*Guide to Hart-Parr, Oliver, and White
   Farm Tractors, A*, Larry Gay
Industrial Heritage, An, Walter F. Peterson
*International Harvester Farm Equipment
   Product History 1831-1985*, Ralph
   Baumheckel and Kent Borghoff
*International Harvester Tractors 1955-
   1985*, Ken Updike
*J.I. Case Agricultural and Construction
   Equipment 1956-1994*, Tom Stonehouse
   and Eldon Brumbaugh
*John Deere Tractors and Equipment 1960-
   1990*, Don McMillan and Roy Harrington
*John Deere Tractors and Equipment1837-
   1959*, Don McMillan and Russell Jones
John Deere's Company, Wayne G. Broehl Jr
Nebraska Tractor Tests Since 1920, C.H.
   Wendel
Oliver Hart-Parr, C.H. Wendel
Proud Heritage of AGCO Tractors, The,
   Norm Swinford
*Ultimate American Farm Tractor Data
   Book Nebraska Test Tractors 1920-
   1960*, Lorry Dunning

Acknowledgements & Picture Credits

The publishers would like to thank the photographers,
manufacturing companies and private collectors who have
cooperated in the production of this book either by
supplying material from their archives or by allowing access
to their collections of vehicles and memorabilia for
photography. In addition to the owners of the Landmark
Machines who are cited on the appropriate pages, they
would like to thank Dick Attaway and Lea McCall of
Caterpillar Inc., Colin Boor, Stuart Gibbard, Keith Goacher,
Lina Landess, Jack Polhill, Randy Nehman, Bonnie Walworth
of Ford Motor Company, and Ray Woolford.

The commissioned photographs of Landmark Machines
and memorabilia and collectibles included in the book were
taken by Neil Sutherland with the exception of the pictures
on pages 116-117 and the main images on pages 176-177
which were supplied by Patrick W. Ertel.

Other color photographs reproduced in the book were
supplied by andmorphoto@aol.com, Patrick W. Ertel and
David Sparrow.

The black and white photographs on pages 40 (bottom
center), 50, 54 and 55 are from the collections of Henry
Ford Museum & Greenfield Village and Ford Motor
Company and are reproduced with their kind permission.
The black and white photograph on page 179 is reproduced
courtesy of Caterpillar Inc.